Narrative Matters

Narrative Matters:
Teaching and Learning History through Story

Grant Bage

First published 1999 by Falmer Press
11 New Fetter Lane, London EC4P 4EE

Simultaneously published in the USA and Canada by Falmer Press
Routledge Inc., 29 West 35th Street, New York, NY 10001

Falmer Press is an imprint of the Taylor & Francis Group

© Bage 1999

First published in 1999

A catalogue record for this book is available from the British Library

ISBN 0 7507 0980 4   cased
ISBN 0 7507 0979 0   paper

Library of Congress Cataloging-in-Publication Data are available on request

Jacket design by Caroline Archer

Typeset in 10/12pt Times by
Graphicraft Limited, Hong Kong

Printed in Great Britain by Biddles Ltd., Guildford and King's Lynn on paper which has a specified pH value on final paper manufacture of not less than 7.5 and is therefore 'acid free'.

Contents

Glossary

AT   Attainment Target (an objective as described in National Curriculum documents, from which teachers plan teaching and assess learning)

DES   Department of Education and Science (government department responsible for education 1964–92 before it became the DFE)

DFE   Department for Education (government department responsible for education 1992–97)

DfEE   Department for Education and Employment (government department responsible for education since 1997)

ERA   Education Reform Act (1988 act resulting in the NC, followed by 1993 equivalent forming SCAA)

GCSE   General Certificate of Secondary Education (exams often marking the end of compulsory education at 16, started in 1988 following the merger of CSE and 'O' levels)

HA   The Historical Association (subject organization to support educational history)

HCP   Humanities Curriculum Project (inter-disciplinary project arising from the schools council curriculum development work started in the late 1960s)

HWG   History Working Group (devised the NC History)

INSET   In-Service Training (professional training for teachers)

KE   Key Elements (1995 NC history term describing what teachers should plan to teach)

KS1   Key Stage One (NC description for 5–7 year olds)

KS2   Key Stage Two (NC description for 7–11 year olds)

KS3   Key Stage Three (NC description for 11–14 year olds)

LEA   Local Education Authority

NC   National Curriculum (result of 1988 ERA)

NCC   National Curriculum Council (government curriculum agency responsible for the NC 1988–93)

NLF   The National Literacy Framework (published by the DfEE 1998 as a detailed guide for schools to teaching the NLS)

NLS   National Literacy Strategy (developed by government 1997–98 to direct literacy teaching in Y1–6)

Ofsted Office For Standards in Education (government body responsible since 1992 for administering and monitoring school inspections)

PGCE   Post-graduate Certificate in Education (generally one-year teaching course leading to a qualification to teach)

QCA   Qualifications and Curriculum Authority (government agency formed in 1997 with overall control of the school curriculum and examinations, taking over from SCAA)

RY   Reception Year

SCAA   Schools Curriculum and Assessment Authority (government agency formed by merger of NCC and SEAC in 1993)

SCHP   Schools Council History Project. The Schools Council was set up in 1962 to advise government and education on the curriculum, with each discipline having separate but sometimes overlapping work. The SCHP did distinctive pioneering work in history curriculum development, partly through developing its own public exam courses and curricula.

SEAC   Schools Examination and Assessment Council (government body formed in 1988, amalgamated with NCC to form SCAA in 1993)

SHP   Schools History Project (independent body formed after the government discontinued funding the SCHP, in order to support schools, teachers and students who were working in curricula developed by the SCHP)

SoA   Statements of Attainment (assessment objective in 1991 NC History)

SU   Study Unit (or History Study Unit, HSU — divisions of content in NC history)

Y1 etc.   Year 1 (NC abbreviation for age range of children, e.g. Y1 is 5–6, Y2 6–7, etc.)

Preface

Teaching has so many starting points and inspirations that a book about pedagogy can acknowledge only a fraction of its debts. The most obvious are to the children, teachers, parents and schools I write about, but my debt is as great to the teaching colleagues, course-goers and pupils who, as fellow educators over the last 17 years, originated or have been critical friends to the ideas and practices I describe.

Current and recent colleagues have similarly stimulated many of the questions to which this book offers some answers. Colin Conner, Christine Counsell, Peter Cunningham, Marion Dadds, Mary Jane Drummond, Michael Fielding, Phil Gardner, Susan Hart, Bob Lister, Mary James, Donald McIntyre, Judy Sebba and Geoff Southworth have, through their talk and teaching, shaped my thinking. John Fines remains more of an inspiration than he will ever acknowledge. Fiona Collins, Hugh Lupton and Rachael Sutherland are the most stimulating of people to work with on storytelling projects. Rita Harvey, Mavis Robinson and Peggy Nunn have helped turn typing into text, and the library staff at Cambridge University and the University of East Anglia supplied numerous readings from which ideas have been born, or stolen. Books are authored by environments, as well as individuals. I have been lucky with mine; long may they thrive.

There are also particular individuals who, during this book's gestation, influenced me deeply. Bev Labbett supervised the PhD thesis from which it grew, and has been a bubbling source of original ideas and inspiring encouragement. Chris Husbands similarly spurred me on. My erstwhile colleagues in Suffolk LEA created a humanities advisory team of which I was privileged to be a member: John Clark, Martin Oldfield, Jill Penrose, Clive Paine, John Fairclough, Chris Garnett, Derek Merrill — and numerous other colleagues from disciplines such as drama, art, English and information technology.

Finally, thank you to my family. To a mother and father who fostered a deep love of history alongside a fascination with stories, and to Teresa, Thomas and Anna. Given my frequent bookish exiles from family life, you have every reason to detest this book. Instead your love is inscribed on every page, albeit in invisible ink.

Chapter 1

The Story of this Book

A history of the world, yes. And in the process, my own. The Life and Times of Claudia H. The bit of the twentieth century to which I've been shackled . . . Let me contemplate myself within my context: everything and nothing. The history of the world as selected by Claudia: fact and fiction, myth and evidence, images and documents. (Lively, 1987, p.1)

Foreword

Since you are reading these words you may already be questioning what history and stories can offer to learning, curious about them *as education*. Suffusing your reading are personal experiences of and public debates about schooling, from which we construct our own stories and histories *of education*. This book weaves both threads to concern itself with history and stories *in education*. In education, history and stories exist simultaneously as curricular artefacts but also as individuals' lived experiences. Public and private, professional and personal, for teaching and learning: we shall be exploring histories and stories as, of and therefore in educations plural. So, be forewarned. This work will attempt to persuade you, through evidence and rhetoric, that stories and histories are practical methods to teach *through*, as well as fundamental ideas to become cultured *with*.

I am now a hostage to your fortunes with this book. I dare not compute the years it has cost, and in minutes you may toss it aside. Nevertheless your reading instantaneously designates this a joint venture in which I challenge you to participate. You cannot agree with everything I say, and despite my admission of advocacy I harbour no ambition to make you. My only desire is that this book will stimulate you to think again about its and your own histories and stories. As a starter I can supply as evidence descriptions of people and events, some settings for context and discussion of the roles of author and genre. I can even offer endings or storylines to assist your reading. Since it is you, though, who decides the meaning of what you read, the resolution of our enterprise now lies in your hands.

Descriptions of People and Events

Empirically, most educational evidence derives from descriptions and interpretations of people and events, viewed through experience and theory. Hexter (1972) sketched a similarly simple code for historical evidence in which the *second record* is our own internalized experience, through the eyes of which we interpret the external *first record* evidence — eyewitness accounts, bureaucratic records, etc. Much of this book's data started life as first record empirical evidence created by the author through:

- interviewing teachers, other educators and children;
- tape recording lessons led and shared — 'participant observation';
- observing and discussing teaching in schools, lessons and courses;
- observing and discussing learning with children in lessons and schools;
- collecting outcomes from children's learning, especially talk and writings;
- a journal of such observations and field notes.

These were complemented by first record theoretical evidence:

- critical reading of educational, historical, philosophical and literary texts;
- analysis of official documents and reports;
- quotations from media sources including classroom materials.

Relations between empirical and theoretical evidence and first and second records are more problematic than this suggests. My participation in and recording of educational events soon, sometimes even simultaneously, entwined first record inter-pretations with second record prejudices. To help account for the practical theory-building that resulted I also therefore open my second record through:

- autobiographical explorations;
- edited journal entries and reflections;
- critical self-analyses;
- reflexive commentaries upon the first and second records.

Brookfield (1995) and Shulman (1987) described similar sources, the former emphas-izing those internal to the teacher or classroom, the latter those external. All will be drawn on in forthcoming pages, spiced by my own prejudice that evidence is to argument what description is to prose: something that convinces the reader not so much of the writing's truth as of its trustworthiness and authenticity.

In creating my first record I have been methodologically eclectic for several reasons. Firstly, this book often describes classrooms I have taught in, and I have had to use the mixed bag of research tools within reach of an unfunded, practising teacher. Secondly, a singular educational research method can dominate or even manipulate data (Stubbs, 1983). Thirdly, there is no 'single conceptual framework or adequate shared vocabulary for describing classroom events and processes . . .' (Edwards and Westgate, 1987, p.52). Lastly, history as a discipline has eclectically influenced the educational, ethnographic methodologies to which I have been most attracted (Hitchcock and Hughes, 1989). As Fischer argued:

> No method exists independently of an object. None can be vindicated except in its application; none can be proclaimed to the world as The Method; and none is other than a useful tool. No historical method is in any sense an alternative to heavy

labour in historical sources. None can serve as a substitute for creativity. (1970, p.xxi)

My 'heavy labour in the [educational] sources' used five ethnographic strategies of naturalistic observation, teaching or helping, eavesdropping, observing in everyday work, and observing over an extended period of time (Edwards and Westgate, 1987, pp.74–5). Major elements of this book's first record evidence also derive from workaday experiences of teaching in scores of courses, schools and classrooms alongside thousands of teachers and children for which no formal or particular records exist. I am indebted to them and have, wherever practicable, offered this text for their scrutiny. Alongside the better-documented aspects of my second and first records, information from tactics akin to informal mass-observation (Calder and Sheridan, 1985) helped shape the tacit pedagogic knowledge that this book then explicitly remakes.

How can we understand such tacit pedagogic knowledge? Some argue that as teachers we often describe our work without any 'explicit theory of learning' (Cortazzi, 1991, pp.132–3). Yet, since the stuff of pedagogy is to present, formulate and invoke interpretations, teachers are not 'passive providers of data' but 'active agents with an interest in what appears for the record' (Denscombe, 1984, p.115). My own research practice reflected such complexities. As in pedagogy, I cared about the record — initially whether a PhD would be awarded and now whether you as reader will enjoy this book. As in pedagogy, research was influenced by current consuming demands and interests as well as optimistic and predetermined plans. As in pedagogy I adopted various identities: sometimes creating data for later interpretation as an historian, sometimes interpreting it in the field like an anthropologist, and on numerous occasions collecting it for no obvious reason at all, in the manner of a manic antiquarian. As in pedagogy I also remain uncertain about how to interpret my multiple participations in many events that have become data. I have imagined, planned, taught, discussed, observed, shared, analysed, recorded, criticized and written up much of the teaching in this book, yet the precise significance of each element remains tantalizingly variable. Approaching reflection and research as a story has helped me navigate such shifting sands.

Settings and Contexts

Enough of descriptions (or data-types). Alone they say little about the story to come. What of the settings in which they were created? It is worth the reader knowing that most of the first record for this book derives from the period 1989–98. During this time I was successively: a class teacher for Y6 children, an advisory teacher in history teaching across the age range 5–16 and now a university teacher and researcher. Previous to that, my intellectual second record had also been informed by teaching most subjects in 9–13 middle schools, but especially English; by taking a couple of history degrees and working in a fair range of locations beyond formal schooling and the education sector. Therefore, practically a decade

3

of storied pedagogy preceded this book's birth in February 1989 as a research project on teaching through oral history. Because oral history interviews seemed to bristle with narrative I started collecting wider evidence about related activities such as storytelling, educational drama and role playing in history. Simultaneously the government was also shifting its attitudes towards school history, via the NC. I was appointed as an LEA advisory teacher for history in 1990. Subsequently teaching NC history in scores of schools, and preparing support materials and running courses for perhaps thousands of teachers has enabled me to develop and test ideas about story and history up to the present day. These years have also seen education and pedagogy embroiled in political debate and media controversy. My eclectic evidence base, the sometimes anecdotal, autobiographical style of writing and the self-criticism of the research questions, is a conscious attempt to contribute to this debate not just by improving teaching, but by bridging academic research with practical pedagogy. The book does not claim to be a true story but does aspire to be a real and critical story about teaching in England in the 1980s and 1990s.

Keeping Critical

Keeping critical is hard work. A recent example chronicled teachers and kindergarten-aged children learning through family as well as fictional stories: 'we need real people, real family members to tell us this part of who we are' (Paley, 1997, p.85). Adult educators similarly need stories to belong to, as my own 1991 research note suggested:

> Research is in many ways about power . . . and freedom. My particular research story acknowledges this power relationship. I am one teacher in a small county in a small country in a privileged corner of a chaotic world. I chose to teach because of my possession of a set of perceived democratic and liberal values . . . In turn these work on and are worked upon by government actions and ideas which need examination at the least, and perhaps critical resistance.

Some claim that teachers naturally synthesize analysis with anecdote (Cortazzi, 1991, p.132), others that:

> One does not have to spend a long time in the teacher's lounge to hear them swap stories . . . about cases . . . a difficult child, a good class, maths materials . . . Case knowledge is characteristic of practitioners who work with people: teachers, lawyers, physicians. (Gudmundsdottir, 1991, p.211)

Even so, it is not necessarily easy for such stories to maintain a critical stance on their contexts. Fitting anecdotes satisfactorily into stories for improvement or political meta-narratives requires a confident and self-monitoring profession: 'A research tradition which is accessible to teachers and which feeds teaching must be created if education is to be significantly improved' (Stenhouse, 1975, p.165). Such a tradition seems essential given increasing governmental control over curricular and inspection systems; stories as research can assist. My own autobiographical record reflects a perhaps typically uneasy compromise between resisting and critically mediating torrents of educational change.

Journal, April 1990

I know there is excellence as well as idiocy in the previous practice of schools and present reforms. My task is to sift these through and build on what I perceive . . . works well with children. One side of me says 'fight back in a struggle' . . . the other 'you win and children win by taking the best bits from everybody', regardless of source.

Although authoring such stories may have helped my pedagogy, long experience suggests that simultaneous research and teaching can be daunting. The key often turns with a few words, lessons and ideas of special significance to us. For me there have been two: talk and story.

Journal, 1992

If I ask what is and has effective history teaching been for *me*, I pretty soon come up with a one word answer: **talk** . . . the potential of student talk, the power of teacher talk and the relationship between the two. This is especially problematic in history, with tensions between curricula claiming that students develop understanding through their *own talk* about historical problems . . . and that history transmits knowledge relying heavily upon *teacher talk* . . . I wish both to shift the ground and introduce my own evidence to make better personal sense of teaching and advising . . . And although I have long practised it, I am only recently articulating how the teaching I enjoy most is when I am telling or empowering the telling of **stories**. By 'enjoy' . . . I mean that I find it professionally and personally satisfying . . . effective in a manner that distinguishes it from other classroom activities.

Keeping critical can start by reflecting upon such experience and is as necessary for policy-makers as practitioners. Consider this observation on a previous curriculum reform, in which pedagogy was sketched as a tradition supported by a wide range of rational and irrational arguments:

> Traditional teaching patterns have not and will not be changed by exhortation or by new materials . . . What is required is to begin a process of change, involving teachers themselves in questioning current practice and drawing up and supporting new definitions of practice. (Goodson, 1978, p.47)

Are times so changed that we can afford to ignore such experience? We may be better advised to keep sharply critical of current curriculum projects, especially those led by central authorities with massive political stakes in their supposed success and effectiveness.

Authors

Following Stenhouse (1978) and the principle that, like history, educational theory should reveal its sources should I now offer readers the personal archive through which I have swum? In all probability you might drown in it, and anyway a word limit of 70,000 curtails this ambition. I shall therefore modify it. Having taken thousands of pages of notes from books and articles, talked to and taped scores of colleagues at all levels in the history community, taught or witnessed hundreds of lessons of history — not to mention concurrently lived a full life as husband, father and friend — I shall now try to explain some of the authorial grounds upon which I have chosen what seems useful — and significant — to share with readers. *You* therefore need to know a little of *me*: not because I am particularly important or different, but because 'In understanding something so intensely personal as teaching it is critical that we know about the person the teacher is' (Goodson, 1984, p.141). Equally, reading this book may stimulate your desire more practically to understand your own teaching, learning and thinking. If so, how could we possibly avoid the *I*?

> When we teach, we think about our teaching . . . Furthermore, it is impossible, when people think, to do so without reference to themselves as people. All the things they know, have studied and have done are present in their thoughts . . . Thus both action and prior experience compose our practical thinking. (Clandinin and Connelly, 1987, p.494)

Aspects of my professional personality will therefore be revealed in this text, and because as readers and writers we now know we are jointly interested in narrative, some revelations will be in storied form. Read these as you choose. Unlike traditional autobiography and biography such research-led stories do 'not assume the authoritative and omniscient narrator who tells the whole truth and nothing but the truth; the teller's perspective is assumed to be a partial one, honestly assumed, and thus open to criticism' (Elbaz, 1991, p.6). Though such approaches locate my work within the traditions of critical or action research (e.g. Brookfield, 1995; Carr and Kemmis, 1986; Elliot, 1991; Stenhouse, 1975), location is not warranty. Stenhouse's assumptive identification of 'open teaching' with action research was flawed (Hammersley, 1993, pp.221–2). My own research methodology has not followed a preconceived blueprint against which you could check its progress, but has seemed more often like a retrospective reading of patterns from actions:

> Practice becomes the starting point, the ground for inquiry and not, as 'technical rationality' would have it, a field of theoretical application. One does not apply theory to practice to enhance its improvement, but, instead, works with practitioners to better understand (and enhance) practice. (Connelly and Clandinin, 1986, pp.294–5)

Even this is not straightforward. I share the suspicion that within the centralized curricula that have dominated English education in the 1990s, teacher-research can be 'hijacked . . . to control pupil learning, to produce pre-defined curriculum objectives or targets without any consideration of the ethical dimension of teaching and learning' (Elliott, 1991, p.52). Yet as a recent and national curriculum-practising teacher, my experience also is of developing — perhaps even subverting — imposed national curricula, inspection frameworks and government initiatives from the inside. As a teacher caring for children — and taking his paycheque — I could interpret or alter these, but not simply ignore them. This book shows some of the consequent struggles of attempting to harmonize two perhaps irreconcilable ideas. The first can be termed as Aristotle's *techne*, technical rationality. This is a form of reasoning suited to the making of products and which English National Curriculum (hereafter NC) structures seem to impose upon pedagogy. The second is *phronesis*, practical deliberation to enable somebody to do something well and virtuously. Although they have lately been forced into the first territory, schools may also be excellent sites for communitarian building in the second: 'The school would be seen as the educational counterpart of the Greek polis — an Aristotelian community based on an education in the virtues' (Wilcox, 1997, p.260; see also Elliott, 1991; MacIntyre, 1981).

Given the complexities and interrelatedness of my own roles and purposes during the writing of this book — let alone any of its numerous other characters — a simple list of techniques cannot warrant it. Instead I offer a pragmatic mix of (technical) actions and (phronetic) qualities to which you as reader will have to decide whether the author bears witness. Foremost is the reflexive realization that to choose an area as worthy of educational research is itself to make a moral and theoretical choice. Research views the world through ideas, creating a conceptual as well as personal interpretation of reality (Todd, 1981). Prominent in this book is the idea of pedagogic self-awareness (e.g. Cullingford, 1995), lying alongside the historical principle of public access to the grounds upon which awareness was built: 'Practice will tend to improve when experience is systematically marshalled as history' (Stenhouse, 1978, p.32).

This book therefore offers systematic marshalling of experience not just *as* or *of* history, but *in* history. Historians 'rarely publish practical reviews of their working processes' (Little, 1983, p.30). Nor do many practising teachers have the opportunity to explain how their pedagogy came to be. I try to offer and account for the *I* in history and pedagogy not because what I say is necessarily valid, but because without such an account pedagogical research — and perhaps even educational research — may be invalid. Yet witnessing and participating in innumerable lessons is not a methodology in itself. Participation and observation offer no '. . . direct access to others' experience. It must be treated as a device for simulating that experience and . . . approximate at best' (Hammersley, 1984, p.9). I also frequently include critical examinations of my own practices and experiences: '. . . not to produce objective data, for that is impossible. Rather . . . to produce subjective data whose subjectivity is sufficiently controlled to allow critical scrutiny. The aspiration is to critical inter-subjectivity, not to objectivity' (Stenhouse, 1978, p.33).

Stenhouse (pp.21–2) claimed over simplistically that analytic history opened the processes by which it selected evidence and came to conclusions and narrative history obscured them by simply 'telling the story' with a glossy threading of plot (also O'Dea, 1994, pp.165–9). I share such fears of storied approaches (see especially Chapter 6), but since I am trying to synthesize narrative and analysis — stories and questions — in the teaching of history, it seems congruent to attempt the same synthesis during reflexive research into that practice. Is there any alternative to such an uncomfortably pragmatic compromise? 'We cannot escape from history. We can only speak from within it. But that does not mean that there is no truth ... only that whatever insights we have will always be provisional and capable of being contested' (Josipovici, 1993). This book shows such shifts and includes autobiographical evidence of my own arguments, opinions and emotions. These may fall into traps of wishful thinking, inauthenticity, propagandizing, over-storying and sentimentalizing (O'Dea, 1994) or into further dangers of storied history (Chapter 6). The alternative to admitting that I did and still might make these errors in pedagogy or research is to deny them — to sweep them into a part of my story that you, the reader, may not see. What trust could you then place in me, either as a teacher or writer?

Genres and Style

The disciplines of story have been honed to help make human sense of experience, from the apparently mindless to the incredibly complex. In research terms sometimes meanings are read from a single narrative, at other times various narratives are synthesized to ascertain meaning (Bruner, 1986; Polkinghorne, 1995). This book adopts both approaches but usually the latter. Clandinin and Connelly (1990, p.244) perhaps overstate how such narrative enquiries contrast with formal or technical methodologies by placing untidy, unpredictable *experience* at their centre: 'there is a basic opposition in principle between the holistic ends of narrative enquiry, the reductionist ends of technical rationalist enquiry and abstract ends of formalistic enquiry' (1990, p.245). In this book's reflections upon practice I employed the methods and sources they describe: field notes, journal records, interviews, storytelling, letter writing, autobiographical and biographical writing, class plans, personal philosophies, writing of rules and principles, picturing and metaphors (Connelly and Clandinin, 1990, p.6). Or, put more honestly, I had been working in these ways for four years before retrospectively justifying it with their particular academic label, which I read in 1994. Why? On one level I was almost developing a methodology by instinct, using what I knew but not necessarily articulating how I had come to know it. Polanyi (1983) refers to such knowledge as 'tacit' and Hexter's historical analogy of the 'second record' (1972) has already been discussed. History teachers may be especially prone to deploying such knowledge. John's research portrays history teaching '... not as a generic process but as one crafted on a powerful mixture of personal-professional knowledge ... such knowledge often defies codification and is acquired after years of experience and reflection' (1994a, p.36). Similarly others have found that personalizing knowledge was crucial to effective teachers: 'more successful change occurred where teachers could not only re-interpret curricula to suit classroom and school contingencies but where they could adapt or re-design curriculum materials to meet highly personal and professional interests' (Butt, Townsend and Raymond, 1990, p.256). This may be especially true of teachers' reactions to the English NC (see Croll 1996; Pollard et al., 1994) as I shall discuss later. Yet the personalized scope of this justification for storied research seems a narrow ambition. When might individual stories accrue to help develop future, collective scripts for teaching in general? I wish to participate in positive but dialectic traditions which, although acknowledging the personal, move beyond it:

> Our difficulty in finding a place for tradition in our own conceptualisations of teacher thinking has to do with . . . liberal theories of education . . . the traditional is seen as equivalent to the conservative and the archaic . . . usually seen in a negative light . . . I do not believe we have looked hard enough to find positive traditions informing the work of teachers. (Elbaz, 1991, pp.14–15; also Ben-Peretz, 1995)

Such positive pedagogic traditions seem as vital as they are scarce, partly because educators 'value the new and trendy' often without seeing that it is 'sometimes the old dressed up in new language' (Anderson et al., 1994, p.9). Despite back-to-basics rhetoric, traditional pedagogy sits uneasily with the apparently insatiable educational demands of the modern state. We need a sense of living, critical teacher-traditions but 'Schooling . . . is constituted of certain practices related to a pre-modern tradition . . . within a larger culture dominated by the tradition of modernity' (Wilcox, 1997, p.259). Here I am not just recording my biases about history and stories in education, I am making the case for grounded stories of teaching an integral part of my argument. That argument ends not just with stories, but with improving actions from them. Hence I include sections entitled **In Practice** at the end of each chapter. These embody some of the ideas discussed through examples of working documents, teaching practices or summarized advice. They are offered as bridges to help readers construct their own syntheses of theories and actions.

Stories are also as fashionable as ever. Popularly 'We have come to think that knowing oneself is achieved essentially through narrative . . . especially . . . confessional autobiography' (Casey, 1994). Biography has been similarly and respectfully elevated: 'In this age of the biography, well-written life stories are judged as creative achievements ranking besides the best of novels' (Bowker, 1993). The storied currents of mass, popular culture are reflected in recent academic thinking about education and research (e.g. Brookfield, 1995; Bruner, 1986, 1990; Clandinin and Connelly, 1995; Isenberg and Jalongo, 1995; McEwan and Egan, 1995; Smith and Smith, 1994; Thomas, 1995 to name just a few). My own use of narratively-derived research methods and structures partially reflects stories gathering social and intellectual momentum. It also connects to a long-held belief that a systematic study of pedagogy should lead the politics and research of education rather than vice versa, as currently prevails. For 'To give stories of ordinary teachers equal status on the public agenda with government reports is to transform the very terms of the argument' (Casey, 1990, p.301; also Bolton, 1994). Stories of ordinary teachers do more than just 'tell tales from the other side'. Publicly such testimony is rich because pedagogy is a complex and real story — not disingenuously simple, as so many of our policy-makers and politicians present it. Personally through carefully storied research 'the desirability of a unified self is replaced with the acknowledgement of the multiple and contradictory self' (Norquay, 1990, p.292). Such a multiple and contradictory self makes for an interesting life and an intricate book; especially when that book simultaneously accounts for itself while offering practical pedagogic commentaries. Autobiographical elements help, since my experiences may resonate with readers', but the story is still imprisoned in selection. Because my past is too vast to express and my present is continuous, all that 'we can perhaps obtain is that which, according to the present perspective, appears to be relevant . . .' (Elbaz, 1991, p.17). So, read this book as an articulation of professional memory-developing through scripts. Scripts 'represent a generalization of past experiences and serve as guides for understanding further experiences, and as a basis for appropriate action' (Ben-Peretz, 1995, p.63). I attempt to make these scripts reflexive descriptions evidencing not just the outcomes, but the thinking processes that created them. They mimic Middleton's (1993, p.174) revelatory autobiographical techniques to 'make visible and problematize our own and others' positionings within . . . phenomena that are our objects of inquiry.' Where as reader you spot contradictions, uncertainties or changes of attitude, there may lie imperfect proof of the writer's attempts to 'trip himself up' as Bev Labbett, my own finest teacher, puts it. The methodological point is to log and explore errors. This delays achieving autobiography's seductively 'unified self' (Grumet, 1990). A more prickly narrative path is recommended, for too smooth a story may reflect self-manipulation: 'In life history research we need to take up the contradictions . . . to foster new connections between memory and practice' (Norquay, 1990, pp.294, 297). In texts and lives 'there are multiple possible narratives and threads' (Connelly and Clandinin, 1990, p.245), so entirely predictable research or teaching scripts seem impossible to achieve. My ambition is different: to aid readers, including myself as audience (Bolton, 1994, p.63), to detect the writer's errors. Consequently, do not read this book as a final true result, but as a speculation offering testable, provisional and practical hypotheses.

I could end this section here, tidily and on a functional note. This would be a dangerously smooth narrative (Spence, 1986, p.231) fostering the illusion of inevitable cause and effect. Let me be as spiky a writer as I urge you to be a reader, by repeating that the story told so far represents only one possibility. Narratives can beguile readers into believing that:

> . . . later episodes are causally related to . . . earlier, rather than continuing to spin out of pure possibility. This illusion of causality is created by the essential pastness of a well-strung narrative, reinforced by its closure, by the apparent completeness of its action. (Crites, 1986, p.168)

Similarly, remain alert to the fact that in extracting and analysing talk from transcripts, I have taken the risk of 'so running the data and its interpretation together that the "evidence" . . . can only support the version of reality which is being offered' (Edwards and Westgate, 1987 p.107). Though such problems of selection and contextual knowledge concern oral historians in general (e.g. Elinor, 1992, pp.78–9), it is hard to envisage alternative courses of action. If the 'whole of the evidence' were put before you as reader would its overall significance be clearer? Selection and interpretation are integral to the practices of teaching or research, only becoming dangerous when pursued through inaccessible or unverifiable means. I am trying to reveal such means, but how do I want you to use them? My bias is that you should pursue the following storylines.

Plots, Storylines and Resolutions

I claim no universal truth for my autobiographical commentaries. They have been educationally and tentatively constructed to:

> Reflect deeply on teaching practice, to see it from a variety of perspectives, to uncover and bring to conscious awareness the multiple levels of presuppositions that inform perceptions and which determine (often unconsciously) . . . interpretation(s) of particular situations. (O'Dea, 1994, p.167)

The researcher as interviewer or writer is as open to scrutiny as those being researched. Modern oral history takes this into account: 'Oral history's . . . subjectivity is at once inescapable and crucial to an understanding of the meanings we give our past and present' (Yow, 1994, p.25). Since I want our mutual meaning-making to be participative, I now speculate about ways in which you might as reader treat the forthcoming story. If this were a novel it would be suicidal so to reveal its plots, problems and resolutions. Since it aspires instead to be trustworthy pedagogic research in a narrative but analytically reflexive tradition, this revelation of the author's desired storylines forms a significant part of its evidence-base.

Storyline 1: Describing some Uses of Story

At a minimum this book describes storied attempts to domesticate the wilder parts of school history teaching and enliven the duller ones. Analysing these may help to develop 'practitioner-led theory' (Knight, 1991, p.138) rooted within but still critical of national curriculum pedagogy.

Storyline 2: An Archive, Chronicle or History of National Curricula

In similar vein the reader might see the book as a chronicle and archive for future NC researchers. Some sections might just provide serviceable history, as an interpretative narrative based upon evidence. Underlying this are questions about modernity and tradition. By normal political reckoning, even accounting for the influence of a fabled if not mythical 'educational establishment', Conservative governments from 1979–97 might have produced a NC in which 'old-fashioned narrative' was explicit. I argue that this has not occurred — and that it should.

Storyline 3: Do History Teachers and Students Act out Storied Roles?

Whether history is taught through narrative as an elementary-style definition or drama as a more progressive equivalent (Alexander, 1995, pp.270–71), it seems to this observer that story lies marrowed in the bones of both pedagogy and history. Storied teaching may reflect innate narrative structures in history and education, rather than being as in Bennett's (1976) view the outcome of teachers 'choosing' classroom styles to reflect their ideologies. If the time is ripe for a new accord between a liberal and a subject-led curriculum (Alexander, 1995), is it time to examine whether history teachers and learners may be storytellers and makers whether we *like it or not* and *recognize it or not*?

Storyline 4: Can Story Reconcile Analysis and Transmission in History Education?

At the height of the controversy over the first NC history it was argued that history:

> ... is not a 'knowledge-based subject'; it is an evidence-based discipline ... the word history both refers to the outcomes of study and to the process by which knowledge is produced: the two cannot be separated. Historical knowledge results from the scrutiny of evidence, and is always provisional. (Husbands, 1990)

Some saw in 1991 NC history 'a consensus ... on the essential interaction of process and content, on criteria for defining these, and on the need for continuity and progression in developing and evaluating children's historical understanding' (Cooper, 1992, pp.1–2). This book acknowledges the pragmatic worth of such a resolution. It also explores how my own views of history have developed beyond 'skills' (typically in NC Attainment Targets or Key Elements) applied to 'content' (in NC history study units). Instead, I examine how teachers can use story to help children learn about these simultaneously *and* be inducted into broader ideas of literacy, culture and tradition. More broadly I am exploring the reconciliation of two apparently conflicting educational philosophies. The first asks pupils analytically to theorize through talk (e.g. HCP, 1970) supported by the teacher acting as neutral chairperson. The second asks teachers to act as agents of cultural transmission through teaching NC histories (e.g. DES 1991 and DFE 1995a). Might the *principled* educational use of spoken history stories harness both?

Storyline 5: Might Spoken, Story-based History Help Children 'Act as Historians'?

In my vision of storied history curricula pupils orally, literally and reflexively construct and deconstruct explanatory narratives about the past, derived from evidence. These alter in shape, length and expression from 5 to 16 but consistently have narrative skills, themes and concepts at heart. Pupils might experience a wide range of stories to transmit knowledge as information and be taught how to interpret, deconstruct and doubt such stories to develop knowledge as understanding. More importantly they would be taught and assessed in the construction, expression and interpretation of their own historical stories from evidence (to finish up with a personalized knowledge). Their performance would be assessed against two ideas: 'how well can this child tell historical stories based upon evidence?' and 'how well can this child criticize the historical stories of others?' It is an attempt to reconcile narrative historiography with the skills-dominated educational ambition to 'teach children to act as historians', perhaps reflecting similar shifts in professional historiography (e.g. White, 1987).

Storyline 6: Is this Autobiography Representative of a Narrative Renaissance?

Are we witnessing a renaissance of narrative in education, and might my experiences and research reflect intellectual developments beyond my personal boundaries? Am I typical, in my changing ideas about narrative, of a group of educators, researchers and historians who now find 'skills-based' or 'competency models' of thinking too crude an orthodoxy? As recently as 1990 it was observed that

> For children to actually sit down and listen to a story smacks of chalk and talk . . . there tends to be an assumption . . . in both Primary and Secondary schools, that if children are not handling sources they are not doing history; if they are not busy doing they are not learning . . . story-telling has fallen into some disrepute. (Farmer, 1990, p.18)

Just seven years later, Grainger (1997, p.20) claimed 'the art of oral storytelling is undergoing a remarkable revival in Britain . . . the myths, legends, tales and history of all its people are becoming woven into the contemporary tapestry of storytelling in Britain.' What has been happening here, and can this book help to explain it?

Storyline 7: Research as Story — Pedagogy as I?

This book tries to record, connect and reconcile different views of history, education and story by drawing on evidence from the vastly differing perspectives of children, teachers, educationalists, historians, philosophers, politicians, journalists and popular cultures. It inevitably 'fails' in the positivist sense because it cannot escape the autobiographical *I*. Instead it attempts to account for and harness this *I* with a storied record of an ignorant man's attempt to educate himself; of his errors and of his awareness of some of them. Failure to realize and record these flaws would, I argue, invalidate research results. For unlike the finalities of literary or religious stories, teaching and research represent tentative, compromising stories to be tested. In this they are akin to the processes of the creative artist:

> When it comes to the ... realization of the idea and to transferring it on to paper, then even the artist's plan begins to change under the influence of the execution of his work, which embodies his self-critical corrections and the elimination of errors ... (Popper, 1992)

Storyline 8: A Clarion Call for Sustainable Education?

Cullingford (1995, pp.11–12) summarized effective teachers humanely as possessing integrity, enjoying learning, being organized, communicating interest and using humour. Might such qualities proliferate if teaching was more confident, as a profession and of its profession? A knowing use of story may assist in developing a confident and sustainable educational enterprise, satisfying learners and parents without marginalizing teachers or belittling classroom experience. I invite you as reader to consider this and, if you are an educator, actively to pursue it by testing the ideas in sections at the end of each chapter entitled **In Practice**. These present in condensed form teaching, writing, thinking and reading ideas from and for storied pedagogy.

In Practice (1)

The following advice was first developed to help practising teachers read narrative research more critically, and has been structured to mirror this chapter's headings:

Foreword

Given that most teachers have restricted access to research resources, story's well-grounded structures can provide a peculiarly appropriate and sustainable research technology in busy schools. The following questions are designed to help readers critically to read research stories and if they wish, then to write their own.

1. Has the writer explained to readers the grounds on which the narrative might be judged, e.g. whether it is:

   - trustworthy?
   - plausible?
   - authentic?
   - meaningful?
   - self-critical and self-monitoring?
   - insightful?
   - open to scholastic scrutiny?
   - resonant with readers' experiences?

Descriptions of People and Events

1. Does the research explain how its evidence was influenced and created rather than pretending that the use of research tools was purely technical?
2. How does this narrative's evidence aspire to accuracy, without the story simplistically claiming itself as true?
3. How does the narrative differentiate between knowledge derived from publicly accountable sources susceptible to scholarly criticism — and private perceptions or prejudices bubbling up from the personal?
4. Does the story pay sufficient attention to analysing rather than describing the actions of the central character?
5. Do other characters influence how their stories are structured and told?
6. Does the narrative method account for the awkward evidence it *excludes?*
7. Does the narrative account for the 'envelope of invention, approximation or fantasy which surrounds every life story'? (Peneff, 1990, p.45)

Settings and Contexts

1. To what extent are 'settings' (O'Dea, 1994, p.165) described to help readers contextualize their knowledge and hence trust of a narrative?
2. Are the story's contexts recorded rather than glossed over? (Polkinghorne, 1995)
3. Does the narrative account for its chosen social and political priorities, rather than assuming it presents a balanced viewpoint?
4. Does the narrated knowledge or self-knowledge relate and translate into individual and collective stories of confirmation or change? Should it have to?
5. Does the narrative in question authentically reflect its contexts, ambiguities and confusions as well as achievements and certainties?
6. Stories dissimulate time: how is time referred to and accounted for in this story?

Authors

1. Have the author's values and practices been opened to criticism through the research story — rather than reinforced by an unquestioning narrative?

2. Written narratives have multiple audiences. Firstly, the writer. Secondly, over-readers of different types. Of this latter group some suggest textual alterations during creation (such as critical friends or colleagues). Others read it at a distance (a buying audience, a reviewer). These 'liminal authors' (Thomas, 1995, p.165) represent posterity. Has the storyteller acknowledged their influence and opened it to scrutiny?

3. How has the writer acknowledged that the meanings of narrative evidence may lie as much in readers' interpretations as in authors' conscious intentions?

4. Has the storyteller attempted to move beyond 'constrained consciousness' (Goodson, 1995) through self-criticism and wide reading, rather than promoting the chosen starting point as naturally justifiable?

5. Are central metaphors in the narrative explained rather than taken-for-granted?

6. How Can I Justify Myself, Who Am I, Where Do I Belong? (Graham, 1991). Does the story's autobiographical element analyse rather than rest upon such classic prompting questions?

Genres and Style

The following genres are commonly used, many in this book:

- biography;
- autobiography;
- life history;
- life story;
- oral history;
- interview;
- anecdote;
- journal and diary;
- observation, especially extended ones;
- folklore;
- myth;
- imaginative and fictional writing;
- oral storytelling and listening (e.g. Connelly and Clandinin, 1990; Thomas, 1995).

1. Are a text's research questions best answered by such storied methods and genres or by more quantitative measurement tools (e.g. questionnaires, systematic observation, discourse analysis)?

2. Is the central vocabulary of the story — its major terms, expressions and descriptions of research — freshly defined, rather than taken-for-granted?
3. Educators use many forms of pedagogic narrative: cautionary tales, origin myths, romantic journeys of self-discovery, justificatory speeches to the jury, comical descriptions of classrooms, tragedies of change and anecdotes of action. Has the writer explained his or her chosen forms during the text?
4. Do technical tools such as footnotes, reference systems and bibliography open the narrative to readers' scrutiny rather than sandbag the text (Bassey, 1995)?
5. Within practical constraints, does the final genre or style of the research report derive from the range of evidence used in its construction?
6. By reflecting and elevating the personal and the idiosyncratic, might stories conserve the status quo by fragmenting desires for change into thousands of weak individual narratives — rather than into a forceful meta-narrative? (Goodson, 1995)

Plots, Storylines and Conclusions

1. Does this research story acknowledge that narrative forms may elevate some versions of reality and suppress others? (Goodson, 1995)
2. Is this text a single story that claims itself as true, or a story constructed by analysing many different stories into an argument? (Polkinghorne, 1995)
3. Has the writer discussed the tension between the awkwardness of a research story's evidence and the attractions of a seamless narrative flow?
4. Does the story tie itself to grander narratives, rather than lie isolated?
5. Does the author acknowledge obstacles that rerouted the flow of argument? (O'Dea, 1994)
6. Does the research story hold open different or difficult endings, rather than rushing to happy or premature closure?

Chapter 2

Story in the Bones

I had a direct line to the past through the myths, legends and stories we learnt . . . if we regularly read stories like that more, they would have. (John Englebright, first school headteacher, October 1994)

Introduction: Are Stories Inescapable?

Story is a 'fundamental structure of human experience' (Connelly and Clandinin, 1990, p.2) permeating everyday plans, rituals, dreams, play and language. Culturally it preceded history, often doing its job as well as others. Therefore narrativization is primary rather than derivative, 'so primary that the real wonder is that the historians were so late in discovering it' (Mink, 1981, p.239). This primary nature makes it as much an 'act of mind transferred to art from life' as an aesthetic (Hardy, 1977, p.12). Its significance is also communal:

> Because we live in groups, we need ways of understanding the actions of others. This requires a cognitive analysis of action in its social context . . . In effect, narratives are a solution to a fundamental problem in life . . . creating understandable order in human affairs. (Robinson and Hawpe, 1986, p.112)

The communal and personal significance of story starts young. Listening to the questions children ask about stories it is 'impossible to . . . maintain that children under the age of six or seven are incapable of reasoning deductively' (Donaldson, 1978, pp.55–6). Longitudinal research into early home and school language concluded similarly that since 'storying is the most fundamental way of grappling with new experience, the best path to this achievement is likely . . . to take them through the domain of stories, their own and other people's. Stories provide a major route to understanding' (Wells, 1986, p.206).

Story's significance persists into subjects like history and into secondary schools. 'Underlying pupils' thinking between fourteen and nineteen seems to be the idea of a past which happens in stories. Stories are given in the unfolding of events . . . and . . . as unfolding, if incomplete, stories in the eyes of the agents involved' (Lee, 1991, p.54). Such thinking also extends beyond subjects. Narrative and paradigmatic are two ways of knowing, Bruner argues, complementary but irreducible to one another:

> A good story and a well-formed argument are different natural kinds . . . arguments convince one of their truth, stories of their life-likeness. The one verifies

by eventual appeal to procedures for establishing formal and empirical proof. The other establishes not truth but verisimilitude . . . One mode, the paradigmatic or logico-scientific one, attempts to fulfil the ideal of a formal, mathematical system of description and explanation . . . The imaginative application of the narrative mode leads instead to good stories, gripping drama, believable (though not necessarily 'true') historical accounts. (Bruner, 1986, pp.11–13)

In developments from annals through chronicles to narrative histories these ways of thinking came to live side by side, as in the processes of contemporary science (Bruner, 1986, pp.11–13). History is — or can be — a reflexive, sophisticated synthesis of the two, a 'more profound and more intricate narrative, in which the story consists not in a progressive unfolding of the present, but rather a series of structural reformations . . . not the end of narrative history but the beginning of a new kind of narration' (Fischer, 1970, p.162). This is the phenomenon I am ambitious to achieve: educational history synthesizing analytically argumentative but still artistically captivating stories (see also Shemilt, 1984). Such an enterprise seems essential in an educational world so ambivalent about creativity and so politically dominated by instrumental and reductive targets, plans, numerical evaluations and the other paraphernalia of presumed accountability (see also Egan, 1990).

Is Teaching History a Personal, Emotional and Virtuous Discipline?

If story predicates thinking, emotions underlie teaching. Children learn better from 'sympathic teachers' (Van Manen, 1991, p.98) and teachers with 'integrity' and 'humour' (Cullingford, 1995). Teaching as work orbits emotions (Nias, 1989) as does teacher development (Drummond and McClaughlin, 1994). Curriculum areas such as history can evoke from teachers the most 'passionate of enquiries' (Dadds, 1995). Academically even 'the old Aristotelian distinctions between history and poetry, reason and imagination, are becoming increasingly eroded' (Southgate, 1996, p.122). In other words, teaching history is an emotional matter.

Journal 1990

In an affective as well as a cognitive sense I feel myself to be an historian . . . To me the past is real, relevant and immediate, with its inhabitants retaining some of the rights of the living: to be valued as individuals, to be treated with honesty and respect, and to have their existence rendered worthwhile by refusing to forget the individual truths that they may have struggled all their lives to discover.

My feelings persist and much in post-modern history reinforces the importance of treating them reflexively (Jenkins, 1997). In contrast, national history curricula for children are emotionally mute: I can find no affective words in their pages (e.g. DFE, 1995a). In this they mirror the technical emphasis of what in the 1970s was termed 'new history'. New history's 'great strength from the point of view of scholarly practice — the concern with original documents — is arguably also a disabling weakness. It narrows time-horizons and prohibits grand narrative ...' (Samuel, 1990a). This makes it even more important for children, through pedagogy, to learn explicitly how to enjoy and criticize the affective in historical learning. Although story assists this, aided ably by the NC idea headlined 'interpretations', there is an even more fundamental mystery children should explore: why bother to learn history at all (Bage, 1996)? Shielding behind the NC, it is easy to dodge the question: but if children did not learn history in school, where would they and from whom? Television, games, friends perhaps, and probably adults' own memories. For a 'version of the past — some sort of version — has already affected every child by the time he enters school' (Rogers, 1984, p.21). This is not a bad starting point and children's lived experiences and own ideas are sound educational beginnings. They are not, though, its end-game:

> The reason for teaching history is not that it changes society, but that it changes pupils; it changes what they see in the world, and how they see it ... Learning history is learning 'standards' ... acquiring a disposition to behave in certain ways. Commitment to truth, respect for the past (however strange and unsympathetic it may seem) and impartiality ... If pupils do not learn these, they have not even begun to understand history ... (Lee, 1991, pp.43, 51)

This returns us to a fundamental question: what sort of knowledge can children create from history? The 1991 NC history working group (DES, July 1990b, p.7) offered a useful three-fold definition of historical knowledge as *content* (the past as subject matter), *information* (facts) and *understanding* (facts related to other facts and evidence). Without the third, the first two cannot be truly educational. This is hardly a new idea: 'Facts, discrete facts ... are merely empty emblems of erudition which certify that certain formal pedagogical requirements have been met' (Fischer, 1970, p.311). Although the HWG's three-fold division heeded Fischer's warning my experience speaks to me that in practice, in mind, knowledge as understanding depends upon a fourth source: the emotions. A storied history curriculum can take more explicit account of this than does one devoted only to short-term skills, or knowledge. It offers historical stories as memorable and emotional investments which may — or may not — yield unpredictable returns from a lifelong future.

As an overall educational aim, this idea is well rooted. Teaching's 'fullest ambition' is 'to develop an understanding of the problem of the nature of knowledge through an exploration of the provenance and warrant of the particular knowledge we encounter in our field of study' (Stenhouse, 1979, p.117). The problem has not been this question but our answers to it. For instance, here is a classically planned, centrist and technical justification for the curriculum: '... life in modern societies is so complex that vast amounts of knowledge and understanding are required for effective living ... these cannot simply be acquired in the process of ordinary life ...' (Rogers, 1984, pp.29–30). Such thinking inexorably leads to national curricula, and when these fail to increasingly prescriptive plans for pedagogy (e.g. DfEE, 1998). Such universal legislative plans for mass instruction can obscure the individual interpretations of curricula that *are* education, at least for those learning it:

> An oddity of educational, as distinct from instructional objectives is that their achievement in each individual case may be different . . . each individual child will develop an historical consciousness — when it is achieved at all — by different means, using different facts and concepts in different combinations. (Egan, 1983, p.157)

An alternative to a clockwork curriculum derived from mechanical metaphors is for teachers to become accountable but also artistic storytellers, inducting children into analytic modes of thought via alluring bodies of knowledge. These ideals remain difficult to realize, as my journal recorded:

November 1992

With Y7 we acted out the *content of the evidence* about the Great Pestilence but I still made no attempt to analyse with the children *how that evidence came about*, despite my professed values . . . WHY? Because this seemed too abstract to address in the time available. My instinct told me that the analysis should centre on what happened, through narrative, rather than an examination of 'how we know what we think we know'. In retrospect I perceive I was wrong but perhaps both elements are needed.

It may be difficult to handle evidence and tell a story but as so much of the rest of this book will argue, it is not impossible. Non-technicist and inclusive traditions view history as a universally accessible, everyday intellectual activity based on enquiring, explaining and arguing about stories: 'Nobody thinks historically all the time. But everybody thinks historically much of the time. Each day, every rational being on this planet ask questions about things that actually happened . . .' (Fischer, 1970, p.316; also Hexter, 1972, p.102). These views root from organic metaphors, envisaging education as what individuals retrospectively make from lifelong experiences rather than derive from centralized curricula or objectives. This is problematic for teachers since most of us work in a state system of politically controlled schools 'delivering' compulsory, planned, mass instruction for extrinsic ends. My examination of story is voluntarily spoken from within this bureaucratic system, but I feel with passion that history has an additional, perhaps more important value that is uniquely personal and intrinsic:

> We appropriate our personal past, in fact, out of the future. Even an archaeological excavation, aimed at recovering the past, can obviously not be conducted in the past. It is a project in which the archaeologist engages his future, and the deeper into the past he wishes to go, the longer into the future he must dig. So it is with the recollection of the self out of the past. To become a self is to appropriate a past, and that takes digging. (Crites, 1986, p.164)

This occurs with the youngest of children. During a history lesson on 'past experiences' with R/Y1 children (May, 1992), I recorded Hayley and Hannah. They illustrate Crites, 'to become a self is to appropriate a past, and that takes digging.'

Hannah:	This is when I were a baby.
Hayley:	Hannah, did you come to my party when I was 5? No you didn't.
Hannah:	I haven't been to any of your parties yet. Nor have you been to any of mine. That's my sister.
Hayley:	I doing my birthday, my birthday party. Hannah look at my party. Hannah, look at my birthday cake.
Hannah:	Where's the candles Hayley?
Hayley:	He-are, four-are.
Hannah:	Oh they're sticking out (giggles).
Hayley:	My party's in a train.
Hannah:	What?
Hayley:	My party's in a train at McDonalds.
Hannah:	Oh you mean that train at McDonalds?
Hayley:	Yeah . . . I used to go there a lot . . .

Spoken and anecdotal history is a source of questions and stories to illuminate an individual's present, and help them imagine their future. Here it links with literature and the importance of audience. History is not history until it is *told*, preferably in convincing and engaging ways. For instance and from the other end of the age-scale the 'occurrent temporality' of talking about historical sources such as poetry, music or painting helped American High School students link 'their autobiographies with the biographies of those living before them' (Gabella, 1994, p.158). Similarly Paxton (1997, p.246) found that five out of six history students 'responded positively' to a 'strong narrative voice — a visible author' in their textbooks. More grandly still, conscious and storied historical knowing may help individuals claim a place in society:

> For the story of my life is always embedded in the story of those communities from whom I derive my identity. I am born with a past; and to try to cut myself off from that past, in the individualist mode, is to deform my present relationships. The possession of an historical identity and the possession of a social identity coincide. (MacIntyre, 1981, p.205)

MacIntyre did not argue that we are bound by the traditions of our physical communities or ethnicity. Rather, history as a social activity educationally helps people to construct stories and then decide which to trust. Educational history is unavoidably political, though its emphasis upon evidence helps create critical rather than compliant citizens. Dangerously, history is also a moralizing activity. Not so much in the sense of discovering and disseminating unchanging moral maxims, though enquiry into these may be a part of history (Baldwin, 1994); more that history fosters the virtue of belonging to and being confronted by living traditions.

> This virtue is not to be confused with any form of conservative antiquarianism . . . an adequate sense of tradition manifests itself in a grasp of those future possibilities which the past has made available to the present. Living traditions, just because

they continue a not-yet-completed narrative, confront a future whose determinate and determinable character, so far as it possesses any, derives from the past. (MacIntyre, 1981, p.207)

History aids virtue by making traditions accessible, criticizable and therefore sustainable. As a narrative it is also moral because it supposes endings. 'What other 'ending' could a given sequence of events have than a "moralising" ending? . . . for we cannot say, surely, that any sequence of real events actually comes to an end, that reality itself disappears' (White, 1981, p.22). As the governmental manager of the national curriculum acknowledged, history helps schools to be 'very moral places' (Tate, 1996, p.2, SCAA speech). How? Perhaps through helping children connect their everyday anecdotes with grand narratives, and to see that those grand narratives can be revealed by questioning the everyday. Do we need a complex, technical, measurable curriculum to achieve this? Children bring the seeds of curricula to school themselves, as stories of their own experience:

> Family life and the school yard provide a microcosm of the forces that have shaped the history of the world and its struggle for a civilised life against the ever-present forces of barbarism. The educational task . . . is to extend their concepts from their local experience and attach them to one part of the real history of their world. (Egan, 1989, p.55)

Egan claims that we need to think of teachers as storytellers. I agree but would extend it. Teachers need to be storytellers not just to tell the real and moral stories of our communities, but to enable children to devise their own moral stories based upon evidence and become storytellers in turn.

Might Stories Be Useful for History Teachers and Learners?

For some thinkers pedagogic knowledge is storied by definition: 'The story is not that which links teacher thought and action, for thought and action are not seen as separate domains . . . Rather, story is . . . the landscape within which we live as teachers' (Elbaz, 1991, p.3). Frances Sword, an experienced education officer at Cambridge's Fitzwilliam Museum, described the equivalent for children:

> Children learn about the Ancient Greeks from stories. Stories that the Greeks made up about the Greeks. Myth, fact, legend, it's a mish-mash . . . Because history is stories. It's just versions. So if children make up stories based on evidence, and we do it all the time — OK . . . How we question children is their prime motivation and . . . we can question in a million different ways. (Interview, 1994)

Story prompts children's powers of telling and listening. A recent author cited an inspiringly crafted, substantial retelling of a traditional tale by a 7-year-old with special educational needs, amongst much other persuasive evidence of how young children have 'natural narrative competence, which can become a significant oral bridge to literacy' (Grainger, 1997, pp.13–15, 59). Similarly, reading or hearing history stories can be more authentic and accurate than working through textbooks, as this 12-year-old told me:

> It's better than doing stuff straight out of books . . . because when you're writing
> out of books you don't actually know the real answer. But . . . when you're writing
> a story you can have different things, you can put what you want in and how you
> think it should have been . . . And when you copy, when you've got to write down
> answers you only get one real answer, and in life you can get more than that, more
> than one answer. (Mark, Blackbourne Middle School, 1990)

Despite these recognized assets ambivalence lingers about teaching through story,
perhaps heightened by pressures for quick-fixes to 'literacy' and anxieties about
'coverage'. Story can assist both but its ambitions are wider namely: 'a slow un-
examinable growth of scenes and images in the mind' (Cook, 1976, p.xi). This was
precisely argued over 20 years ago in a study of London primary schoolchildren:

> The stories they hear help them to acquire expectations about what the world is
> like — its vocabulary and syntax as well as its people and places . . . And though
> they will eventually learn that some of the world is only fiction, it is specific
> characters and events which will be rejected; the recurrent patterns of values, the
> stable expectations about the roles and relationships which are part of their culture,
> will remain. (Applebee, 1978, pp.52–3)

Stories socialize children into wider worlds, offering youngsters access to the
values and experiences of their elders. It has been claimed that a major reason for
inner city educational deprivation amongst pre-school children was that as toddlers
they had not heard the 'three thousand to four thousand stories which the educated
middle classes have told their children' (Brighouse, 1994). Such reasoning helps
explain why another researcher, after empirical comparisons of 25 fictional and
non-fictional history texts, argued that the fictional were more educationally accessible. Linguistically they resembled speech and 'the facts conveyed incidentally in
stories are often more memorable than those deliberately set out in textbooks'
(Perera, 1986, p.64). Affectively too, storytelling matters. As a colleague and drama
advisory teacher commented during a history INSET session (April 1992) 'You
actually held the class much more . . . If you make eye contact with the youngsters
whilst you are telling a story, it has much more magic somehow.' Fines echoed the
point: 'Telling . . . is essentially eye to eye . . . many teachers attempt to work entirely
without feedback, attempting somehow to guess the reactions of their students, but
the storyteller may tell it at once' (1975, p.102).

What of storied history beyond schools? All ends of the spectrum concur.
Academically Elton argued that 'the historian must read not only with the analytical
eye of the investigator but also with the comprehensive eye of the storyteller'
(1967, p.109). *Philosophically* Ricoeur claimed 'because most historians have a
poor concept of "event" and even of "narrative" — they consider history to be an
explanatory endeavour that has severed its ties with storytelling' (1981, p.167). But
chronology binds history to narrative. *Alternatively* Foucault forwarded the concept
of 'genealogy' to simplistic linear stories, arguing that self-awareness character-
izes knowing narratives: 'If we make a book which tells of all the others . . . And
if it does not tell its story, what could it possibly be since its objective was to be
a book?' (1977, pp.139–40). *Practically* Hexter argued that the ever-expanding

evidence-base requires arranging, interpreting and communicating, and that stories do that efficiently for history (1972, p.89). Sourced by such scholasticism and the information age, *educationally* the knowledge available to curricula burgeons in all fields (Wragg, 1993, p.2). In the face of an acknowledgedly over-crowded curriculum (Croll, 1996; Pollard et al., 1994) teachers need accountable, verifiable simplificatory systems to enable children to access information: stories perhaps? Three teachers' reactions to a classroom story illustrate this:

> *Teacher 1*: We were absolutely thrilled and amazed weren't we?
>
> *Teacher 2*: Yes . . . even the non-writers really . . . did take off . . . And the details that they remember as well . . . that's all their own.
>
> *Teacher 3*: 4L and the Guy Fawkes story, I did a storytelling technique with them afterwards using what you'd done — brilliant . . . just the amount of highly complex detail they retained and knew . . . (Castle Hill Junior school, November 1991)

Such educational stories have dangers which are discussed in Chapter 6. As I perceived in this instance, story can swamp children so that they reproduce rather than criticize its transmission. Errors like this lead me to argue that, to be counted as educational, stories should slow, freeze, reverse or control their representation of real narratives so that children can think at their own pace. I am not therefore arguing to replace analysis with narrative in any simplistic manner. Rather, I am trying to understand how narrative may be so ingrained throughout history, time and thought that its scrutiny needs planning. Intuitively this is what skilled historians do for 'clock and the calendar provide no guidance to the appropriate dimensions of a historical story . . . The historical storyteller's time is not clock and calendar time; it is historical tempo' (Hexter, 1972, p.225). So it may be for history teaching. We can neither let children work from the historical record as it is, because it is too bulky, nor time as it is, for it is too long. Instead we have to find ways of mediating between actual events over chronological time, *the past*, and the need for the child to interpret records of that past into *history*. The principled use of historical story may provide this mediation, extending control over records and interpretations of historical events to children and teachers. Story has always been used thus: 'The most fateful decisions are inevitably suspended during the course of a story. We know that discourse has the power to arrest the flight of an arrow in a recess of time' (Foucault, 1977, p.53). Discourse also boasts numerous other powers in history, discussed next.

Story, Language and History Teaching

Many have explored the difficulties of historical language (e.g. Curtis, 1994; ILEA, 1979; Perera, 1986; Wilson, 1985; Wishart, 1986). Its apparent ordinariness 'can often create the illusion of coherence and intelligibility' with 'so much about the intentions of the actors and the "meaning" of their actions . . . simply taken for granted' (Edwards, 1978, p.63). Edwards' answer was to construct historical talk as exchange, with teachers offering information and ideas, and pupils talking back of their understandings; monitoring and evaluating, through talk. Historical proper nouns like the Black Death sound 'factual' but contain massive generalizations. These can cause 'the greatest problems, either because they are assumed to be part of pupils' general knowledge (and so to need no teaching at all) or because the teacher acts as though the context-specific meaning will simply be picked up as he goes along' (Edwards, 1978, pp.57–60).

Two real anecdotes illustrate this well. Walking along seaside cliffs during a wet August in 1996, a passing toddler gestured below. 'Castle mummy, castle,' she cried, pointing towards a 15-foot, square and inflated rubber giraffe. Was she wrong? This thing's name in a previous experience was a castle — a bouncy castle. I related this tale to a teacher friend, who matched it. Her toddler's historical Sunday afternoon outing (and theirs) was irremediably marred because the grey, cold ruins refused to bounce up and down, when she had been promised a *castle*. To direct discourse effectively in a discipline, children need to relearn linguistic rules which 'may be very different from those to which they are accustomed. Most disconcertingly . . . they often have to abandon "everyday" interpretations and explanations' (Barnes, 1992, p.91).

Such ambiguity and everydayness in structures like *story* and terms like *castle* can also aid teaching, by simplifying arcane historical knowledge. This education officer described how she enticed children to analyse her museum's Ancient Greek Gallery:

> I talk to the children a great deal about fact and fiction. We talk about evidence — we are going to make up a character, this character is fiction but it is based upon evidence. What evidence? The evidence around us . . . I have woven a story. In this story . . . let's say there's a child whose father is in the Athenian army. We want to find out as much as we can about: journeys, weapons, armour, food. (Frances Sword, February 1991)

Views of story as an enquiry, motivating rich language from direct engagement with interpretations of the past, arise elsewhere (e.g. Fairclough, 1994; Little and John, 1990). Fines also argued that storytelling brings motivational suspense to teaching by treating 'history as a forward moving development . . . rather than a backward-looking analysis or explanation' (1975, p.97). Story concentrates thinking since 'the storyteller is concerned with . . . split-seconds when decisions were taken' (ibid.). Fifteen years later Fines doubted that the 'Vital storytelling tradition which was a part of the English educational system has actually survived the beating which it took in academic circles . . . there's a great anxiety about it' (Interview, October 1990). Such anxieties have been easy to find. Remarking on his peripatetic experience in 20 primary schools, this teacher rarely saw historical storytelling by children: 'Story isn't geography, isn't history' (Peter Box, 1991 interview). Storytelling by teachers hardly ever happened:

> You don't see telling stories . . . Yet . . . that was what has been most influential on me as a person, someone telling me, recounting their own life experiences . . . the oral not the book. Those memories are there now and are very powerful, out of the oral not out of the written. I very seldom see any of that . . . Story is a very powerful medium, so powerful. (ibid)

The causes of such decline were multiple. Fines identified:

> The move to make the classroom into a factory where everybody is doing busy work, and the move towards groupwork . . . there is a great deal of anxiety of looking a goof in front of a class . . . and story is dangerous, it is seductive . . . a moulded vision of the past, it leads you into heroics and personalization of central characters . . . (Fines interview, October 1990)

This swing against storied history was sometimes supported by teachers moving towards enquiries based on evidence, yet the two are not mutually exclusive. The question which both enquiries and stories answer is:

> Did this really happen? . . . this notion of reality is at the heart of it. Evidence itself isn't in the heart of anything, evidence is just a procedure for getting somewhere. The thing that is important is the notion of reality, because the big question is 'why should I do this rather than read a novel'? Reality is the answer to the question. (Fines interview, October 1990)

Fines also echoed Wedgwood's (1960, p.81) assertion that 'Literature and history were joined long since by the powers that shaped the human brain; we cannot put them asunder.' For him 'Historians can't help but work in the story mode because that is the picture frame of history, the format in which it naturally appears' (Fines interview, October 1990). This book attempts critically to test such revisionist ideas, part of a gathering movement supported by empirical classroom observations: 'Many children dislike reading; on the other hand they enjoy story. History teachers, therefore, should have both the confidence and skills necessary to storytell' (Farmer, 1990, pp.18, 19). Drawing upon similar lengthy experience of infant history Cox and Hughes argue for story as a 'shared experience' a 'meaning-ful context for activities' and 'a means of introducing children to . . . the world of the past' (1990, p.3; also Cooper, 1995a, Chapter 2). Fines, Collingwood and I make related but more radical claims, that story is useful to history education because historical knowledge derives from a narrative form.

> The historian and the novelist . . . each makes it his business to construct a picture which is partly a narrative of events, partly a description of situations, exhibition of motives, analysis of characters . . . as works of imagination, the historian's work and the novelist's do not differ. (Collingwood, 1946, pp.245–6)

Such cognitive arguments have been reflected by storytellers (e.g. Maddern, 1992, p.6) and paradoxically, given the NC's nature, surfaced in government KS1 advisory documents. Story '. . . appeals to children's curiosity, emotions and imagina-tion. It is an effective way of extending vocabulary, introducing new knowledge and addressing moral issues' (NCC, 1993, p.33). According to some, storied cognition in history extends to adulthood: 'Underlying pupils' thinking between 14 and 19 seems to be the idea of a past which happens in stories' (Lee, 1991, p.54). If teaching history through telling stories can ease the difficult relationships between history as content, information and understanding, learning history through being asked to make history stories can transform information into meaningful, motivating and therefore educational experiences. The social act of telling stories triggers learning as it requires the ordering of untidy knowledge, often lying around:

> In the form of general and imprecise recollections, scattered and possibly incon-
> sistent pieces of verbal information . . . all of which will be organised, integrated,
> and apprehended as a specific 'set' of events only in and through the very act by
> which we narrate them as such. (Herrnstein-Smith, 1981, pp.225–6)

It is possible not just for teachers to become history storytellers, but children, too: and for isolated storied ideas within NC history to be extended to permeate the whole of the NC history curriculum, textbooks and teaching. An article typic- ally critical of state schools in that newspaper illustrates the opportunities being lost. Adjacent to a headteacher's complaint at having to retrain secondary history teachers to teach literacy skills it opined that older pupils 'can read a simple story but find it difficult, if not impossible, to understand the more complex language of secondary school textbooks' (Sunday Times, 'Lost for Words', 12 June 1994, p.13). So why not rewrite the courses as stories that can be understood and participated in? (See also Perera, 1986, p.53.)

Similarly, whole-class teaching and active learning need not be a contradiction in terms, thanks to storied pedagogic ideas such as 'teacher-in-role' (e.g. Neelands, 1990; Vass, 1993). A former colleague and drama advisory teacher (Rory Kelsey, April 1992) articulated the technique: 'It allows the teacher to take on an information- giving role that doesn't carry the authority of the teacher. So you can be as pro- vocative as you like, because it's not you, you are talking for someone else, you are being somebody else . . .' In practice it is sometimes difficult, as in the following classroom extract (Barnham School, Y2, 1993). During a history-led topic on Early People and after reading and showing the fictional, poetic but often credible book *Mik's Mammoth* (Gerrard, 1990), I asked the children to mime a *frozen picture* of 'early people working' for others to guess at and interpret. Then as 'teacher-in-role' I mimed a specific action from the book (spearing fish) and asked them to deduce exactly what I was hunting. Was the result merely transmissive recall via a guess- ing game? Eight out of the 16 children's suggestions, in three minutes and five seconds of dialogue, were repetitive (ranging around pigs, rabbits and birds) and two more unlikely, if fun ('flower' and 'fly'). Much of the children's theorizing was hardly breaking new intellectual ground. Yet I also argue that this storied contextualization, the repetitions and simple 'teacher in role' helped children to bring prior knowledge and fresh ideas to bear upon the valid historical question, *what did early people hunt?* Where I see such analytical thinking and interactions occurring more strongly, I have italicized the text. At least eight different children spoke, although identification problems lead me to label them only by gender.

Grant:	Sometimes early people did this . . . *I wonder what they were aiming at?*
Girl:	Pig?
Grant:	*Yeah, but would a pig just sit down there?* (points)
Children:	No.
Boy:	*No it would run away.*

Grant:	A pig would run away. *So what am I actually trying to shoot?*
Girl:	Boar!
Boy:	*I know what you're trying to shoot. You're trying to, you've got some-body down you're having a fight with somebody and you've got him down and you've got him by the really really shake shaking the neck.*
Grant:	*But they wouldn't sit still would they and I'm not very strong so . . .* In any case I'll give you a clue.
Boy:	*They would try and get away . . .*
Grant:	You're right. It isn't a person.
Girl:	*They're not actually statues.*
Boy:	They're wild pigs.
Grant:	*But a wild pig would run away wouldn't it?*
Boy:	I know what they would do.
Grant:	But I'll give you a clue. It's an animal or a creature of some sort.
Boy:	A flower? (laughter)
Grant:	*Well I wouldn't shoot a flower. Cor, this wild flower I'm going to shoot it with a bow and arrow.* (laughter, shared joke). Oh it's moving a bit.
Boy:	*Rabbit, rabbit.*
Girl:	*Deer.*
Girl:	A boar.
Grant:	*Well a rabbit wouldn't let me get this close to it would it?* It's only down there.
Boy:	*No that would just run away so fast . . .*
Girl:	*It might be a hare.*
Grant:	Now come on think . . . sst, arr. (mimes).
Boy:	*Bird, bird.*
Grant:	I'll give you a clue. It was in *Mik's Mammoth*. *But a bird wouldn't get this close would it, see look.*
Girl:	Oh, I know!
Grant:	*See where I am.*
Girl:	*Oh a bear.*
Grant:	Well remembered but it's not a bear 'cos a bear would run away *and if a bear was just there and I was here with my arrow he would just turn round and biff me one because bears are pretty fierce.* No this is another creature that's much smaller than a bear. No, don't do that.
Boy:	*I know, is it a fly or something like that?* (laughter)
Grant:	*I may be an early person but I don't eat flies, ughhh . . .* (laughter). Let me give you a clue, it's bigger than a fly but smaller than a bear.
Girl:	*A bird.*
Grant:	No, I think I've got you lot this time . . .
Boy:	*A tortoise! To make tortoise soup!* (laughter)
Grant:	*Yeah, a tortoise now that doesn't move very quickly. Yes, not bad. Not a tortoise because we didn't have tortoises in* Mik's Mammoth *did we?*
Boy:	*And the spear would just bounce off his shield.*
Boy:	*Shooting down a rabbit hole?*
Grant:	*Now that's a good idea but we didn't have any rabbits in* Mik's Mammoth *either.* Remember there were, there was something that Mik ate.
Girl:	Fish!
Girl:	Fish

Grant:	You clever girl. Yes . . . it was fish. Early people used to use bows and arrows and spears not just to hunt big things but also to hunt fish.
Girl:	*How would they get the fish?*
Grant:	*Well they would stand by the river bank and make sure the fish couldn't see them . . .*
Boy:	*And they would fire . . .*
Grant:	*And they would fire straight into the water.* OK we've got one more to see before it's breaktime . . . (ibid)

I cite this short and unremarkable classroom episode because it is just that. Girls fight to make their voices heard and the teacher struggles to manage multiple responses. These link to my argument that teaching remains a momentary but sophisticated and dilemma-ridden activity, not easily amenable to control by theorizing or legislation. Story here helps answer many such persistent teaching dilemmas: balancing *centralization* (teacher planning in advance, transmitting content by telling story, directing questions, controlling classroom) and *decentralization* (children constructing their own interpretations of transmitted content, performing, speculating, listening, suggesting, discussing). The often messy and eclectic-looking results do not necessarily reflect muddled teacher behaviour. An apparently simple 'telling' might actually be the skilled, flexible and intuitive combining of apparently divided educational traditions: for instance, the traditional and progressive (Bennett, 1976) or elementary and developmental (e.g. Alexander, 1995; Pollard et al., 1994). Forthcoming chapters explore whether such apparently differing traditions have longer roots, exploring teaching-as-telling in the Christian tradition, and teaching-as-asking in the Socratic mould. They also question the apparent necessity and usefulness of such divisions. Might teacher as storyteller *and* prompter be a model which, if principled, helps teachers transmit cultural treasures as well as guiding children to interpret them? Furthermore, evidential and principled storying may help reconcile 'both experimental theory and naturalistic empirical enquiry with the common sense craft knowledge of the expert teacher' (Galton, 1995, p.106). We shall see . . .

In Practice (2)

TEACHING-AS-STORY-TELLING

The T-A-S-TE Project researches & tells local-global stories for teaching & learning

Our experience is that stories underpin educational practice because:

1. **Most children use and enjoy stories.** *Everyday culture —* conversations, spoken or written literature, toys, playground

imaginings, television programmes and computer games often use narrative forms and features. Teachers can build upon these to make *specialized cultures* more accessible.

2. **Stories put meaning centre-stage.** Effective education makes itself meaningful to learners and in turn helps them to make meaning of their world.

3. **Stories inspire curiosity** by exploring the unknown or unexpected. Stories can domesticate experience without dulling it — providing mind-sized portions of content, puzzlement, evidence, explanation, wisdom etc.

4. **Literacy relies upon stories of fact and fiction — or information and imagination.** Rich experience of story lays the foundation for literacy and therefore for success in at least 2 of the '3 Rs'.

5. **People use stories** *to think with.* Storied structures (e.g. plot, motive, character, resolution) help children and teachers to reflect upon and order many aspects of experience — across and beyond the school curriculum.

6. **Stories hold up mirrors.** They offer role models and test cases to explore: how might *we* behave in a story's real or imaginary situation?

7. **Some stories ignore labels.** Children labelled as 'failing' or 'difficult' can respond to story with unexpected interest and concentration. Powerful stories are educationally efficient because they stimulate attention and mobilize cognition.

8. **Stories absorb and stretch agile minds.** For instance a family history, stamp-collecting, ornithology, sport, theatre — all offer stories of sorts that can sustain lifelong interest. Fine stories bear repetition and revisiting: they even improve with it.

9. **Stories communicate information.** If well-told and actively-experienced stories encapsulate and transmit enormous amounts of data (e.g. scientific, historical, religious, technical) in memorable chunks.

10. **Stories suffuse curricula.** Narrative is present in most curriculum areas (especially English, History, Geography, Science and RE) as well as in the lives of families, friendships, staffrooms and playgrounds.

11. **Stories can simplify.** Storied themes are especially useful in making bulky and arcane National Curricula or Schemes of Work appealing and accessible to teachers and children.

12. **Stories are flexible.** This makes them useful as curriculum organizing devices because teachers and children can influence their duration, content, challenge and interactivity.

13. **Stories change people.** Powerful stories can transform individual children and teachers by altering understandings and behaviours — temporarily or even permanently.

14. **Stories explain and moralize.** Since stories rely upon beginnings, actions and endings they explain how and why things happen. Familiarity with these processes is a fundamental aim of most curriculum systems and subjects.

15. **Stories initiate people.** Families, teachers and schools induct children into society through stories — of belief, of science, of the past — because some of these systems' most important outcomes are stories (e.g. in the bible or Koran, of evolution or fertility, of community identities or origins).

16. **Stories explore people.** Narrative forms such as anecdote, autobiography, comedy, myth or tragedy traditionally describe and analyse human motivations in rich detail: and education relies upon human understandings and inter-actions.

17. **Stories analyse consciousness.** Teaching and learning are lived, rather than recorded experiences relying upon specialized, often acute sensitivities. Narrative's expertise in describing and explaining events as though they were happening in the present contributes greatly to accurate and authentic analyses of educational processes.

18. **Stories individualize experience.** All experiences have a local or individual dimension and education is particularly influenced by individuals and contexts. Rich narratives can value local cultures and ideas simultaneously with opening their contextual idiosyncrasy to comparison and scrutiny.

19. **Stories are educationally ancient**. Story's educative traditions appear well before the 19th century industrialization of schooling. In an age when 'industry' as organization or metaphor is under widespread stress and criticism, pre-industrial teaching and learning methods may reward closer examination.

20. **Teachers already use stories.** Story is already skilfully deployed by many teachers, who can build on existing expertise in the quest for improvement.

Don't take our word for it — please test these claims against your own experience or practice.

Chapter 3

Making Meaning of Story and History

For children history is a story, and a story mainly about people . . . History's contribution to a humanities curriculum is not a body of facts and concepts . . . a mass of knowledge which excludes the one knowledge that Coleridge thought it was every man's duty and interest to acquire — self-knowledge. (Schools Council, 1969, pp.12, 15)

Beliefs from the Author's Second Record

History is a negotiated practice offering provisional knowledge. This necessary provisionality does not eliminate the possibility of 'some truths prevailing for centuries, perhaps forever. And one of the responsibilities of history is to record both the survival and reformulation of old truths' (Appleby, Hunt and Jacob, 1994, p.284). Story's special relationship with education seems to me to be one such 'old truth' still useful to modern teachers. Even though contemporary practice offers 'models of teaching in abundance' (Joyce, Calhoun and Hopkins, 1997, p.25), it is tempting to think that the oldest may be the best. To stories I would add playing, observing, questioning and talking as equally attractive and anciently grounded learning and teaching strategies. This text frames them within stories rather than vice versa not because they are inferior, but to promote story as a particularly sustainable and humane educational tool. Story also permeates history: 'In historical discourse, the narrative serves to transform into a story a list of historical events that would otherwise be only a chronicle' (White, 1987, p.43). I therefore offer you a definition:

History is the construction and deconstruction of explanatory narratives about the past, derived from evidence and in answer to questions. This can be explained to children as finding answers to questions and questions to answer, by taking apart and putting together again real stories about the past.

If the story is now complete, why read on? Let me explain. I am not writing to make you follow my lead, just as I try not to teach history lessons that *tell* the truth. I am writing this book — and I try to teach history — to a democratic design. That design offers testable interpretations not *as* truth, but *of* truth. On such grounds I now offer a research note about which I have been arguing with myself for a decade.

Why am I attracted to teaching history through story?

1. In my teaching I use narrative as bricks and story as design. Temporally, narrative connects different episodes which become a storied whole through imaginative reconstruction. Communicatively, turning narrative episodes into open stories publicizes historical understanding: storytelling not just in narrow literary senses, but as an educational conversation. Pedagogic rather than academic history has taught me these things.

2. But what if I, or the curriculum, was in error? To avoid imprisoning learners in my own understanding, I must extend their experience of and power over narrative . . . a main teaching task is to mould NC history and central curricula so that they can help children to criticize, create and enter historical narratives. In this light story has much to offer bureaucratic curricula: to recast them in forms that are attractive and accessible to learners.

3. Despite marginal changes, in my everyday classroom experience narrative still tends to be identified with fiction (e.g. 'write me a story') or with conservative written pedagogy (e.g. 'write me an essay'). There is a much broader assumption worth testing: can we teach history or any other discipline without teaching how to think, listen, talk . . . and then write in an explicitly narratological way? Talking as well as writing ourselves into reasonable stories seems especially important because:

   - Storied talking and listening precede extended writing but help develop it.
   - Spoken stories reflect audience reaction more immediately than writing. They require instantaneous rethinking and redrafting . . . higher quality language can result. Listeners heckle, fall asleep or change the subject in ways that are painfully but usefully obvious to tellers.
   - Talk's tentativeness is of special value in the historical process, since it allows for easier speculation upon the many meanings of evidence.
   - Spoken stories are already common in classrooms, with many variants between the relatively well-known extremes of pedagogic formal lecture and incidental anecdote. Consciousness and control of these may enhance our pedagogic craft knowledge (e.g. Jackson, 1995).
   - Much history derives from talk, written down . . . Analysis of evidence should attempt to reconstruct or consider the talk and stories surrounding its creation (e.g. Domesday book, church buildings, film as source or experience).
   - Spoken stories have a longer educational lineage than their parvenu literary cousin. Can the pedagogic 'problem' of transmitting knowledge to create understanding have changed much since the first verbal transactions between people — whether on the plains of Africa, or elsewhere? Orality prefigured literacy (Meek, 1991, p.14) and our pedagogic

past may shed light — or doubt — on what we think we know about teaching today.

- Spoken stories have a psychological lineage: is the development of thinking not just reliant upon but inseparable from narrative and talk? At the least it seems likely that story teaches tools for thinking — if not thinking itself.

4. That's all very well but . . . Narratives have historically been used to symbolize exclusions from, as well as invitations *to* power. History is littered with pulpits, parliaments and pedagogies run for the benefit of those *already* in control. Even when their stories have been interrupted or conversational — argumentative or dialectic — their association with power is undiminished. A curriculum to decentralize control over knowledge may depend upon exposing the powers which spoken and written narratives confer. If to control narratives is to gain power, should all disciplines more consciously develop powers of constructing narratives (written and spoken) and evaluating them (listening, analysing, criticizing)?

5. Ethics and civics intrude, since all societies host competing and explanatory stories: faiths, advertisements, political broadcasts, advice lines, reminiscences, lessons etc. Many explain our everyday, personal histories. If we do not fully understand their strengths, we may fall prey to those who do: witness the persistent, global influence of news broadcasters, ethnic orators or advertisement-led consumerism.

6. Although belief systems often express themselves in stories (e.g. Christian parables or Norse sagas) one does not have to believe in these systems to value them. If certain stories contain the distilled practical or moral wisdom of millennia it would seem perverse not to retell and criticize them in schools: historically, or morally.

7. Liberal, democratic ideals concerning voice are relevant. Perhaps it is democratically essential for citizens to develop individual historical voices? Voice is typically important in teaching English, and often commented upon in reports. Yet a history teacher who wrote, 'I'm so glad with the way Tina is expressing herself through her ideas about history', would be unusual. We still seem unsure of how to value children's historical theories: might perceiving and telling them as stories assist?

8. The affective is also significant. Stories can communicate emotions in powerful ways. Emotions matter in history, and if well-used, narratives can improve learner motivation.

9. This affective element extends to teachers. Have we been seduced by history stories? On one side lies fear of story's power to distort and persuade; on the other a mythical golden age where learning was easy, story pre-eminent and wickedly compulsory and industrialized education had not been inflicted upon innocents (e.g. Alexander, 1995, Chapter 6).

10. Such ideas suggest to me that:

- If narrative is central to history then telling stories is central to history teaching — as is initiating learners into the storied roles of listening, recording, making, interpreting, prompting and telling stories themselves.

> • Might models of teaching-as-storytelling challenge the public and often private self-image of the modern teacher as deliverer of a nationalized, technicist and pre-packaged curriculum — rather than an autonomous, creative and morally responsible leader?

My assumptions and hopes about teaching, history and story are being exposed to arm the reader with my thinking and protect them from my prejudice. This approach is a model for pedagogy as well as a communicative device. It acknowledges the importance and the impossibility of attempting universal or timeless definitions. It also pursues 'old truths', concurrent with admitting that questions about school history stories soon become qualified by our cultural context: what does story mean *now*, educationally, in the new millennium?

What Is a School History Story?

Stories may comprise 'events, characters and settings' (Toolan, 1988, p.12), but since such materials are the building blocks of most human endeavours, including history and education, this compositional definition does not move us far. More interestingly educational stories such as those offered by history seem to blend factual with fictional aspirations to truthfulness, laying storied and often oral forms over the already-narrativized 'real' life. The implication is that although sophisticated analytical language can be taught through history (Counsell, 1997), this can grow naturally *from* stories rather than be juxtaposed simplistically *against* them. I am making the argument that story is not a form that children grow out of, rather something that adults grow into in ever more complex and reasonable forms — of which one is history.

Wray and Lewis (1997) enumerated non-fiction genres and described practices ('frames') to help children develop such writing. The model of story I offered earlier assists these ends, by exploring the complex relationships between fact and fiction. Although most texts and genres could be singularized as one or the other, my experience emphasizes fact and fiction's frequent overlapping. A discipline like history offers the chance to speak and listen, then write and read texts that are simultaneously narrative and analytical, imagined and real. This may help children learn a 'broad definition of literacy' (DfEE, 1997a, p.45) that blends rather than stereotypes fact and fiction, or as I prefer 'information and imagination'. For instance the common storied forms below can be predominantly information-rich or imaginative according to authors' purposes and audiences' expectations. Universal genres are identifiable, but the truth claims attached to them defy simple generalization. How can we tell whether Wilfred Owen's poetry is a truer record of the Great War than a diary, graveyard or film? Or truly test an echoing literary claim that 'First-person narrators can't die, so long as we keep telling the story of our own lives we're safe' (Barker, 1995, p.115; Prior's diary). Truth here depends more upon tellers' and audiences' reflexive awareness of the kind of story each is trying to tell, less upon its genre. History stories help learners grapple with this fundamental cultural idea, synthesizing like many great disciplines (e.g. science, music, art) information and fact with imagination and fiction.

Some Common Storied Genres Linking Information and Imagination

- autobiographies
- biographies
- court proceedings and cases
- descriptions
- diaries
- educational and performance drama
- essays or elements of them
- explanations of events
- film — especially imaginative but also persuasive forms
- folklore and folk tales
- games and simulations
- legends
- letters
- life stories
- media reports and representations
- memories
- monologues
- museum displays
- myth
- narrative visual artforms — e.g. some paintings, tapestries, murals, etc.
- oral histories and presentations
- personal anecdotes
- poetry
- procedural descriptions
- recounts of events in the past
- reminiscences
- reports of events in the present
- sayings
- songs
- television — especially drama, news, investigative and documentary programmes
- titles, terms and pronouns
- written fiction — stories and novels

Although these genres clothe educational story, they tantalizingly fail to embody it. It remains difficult precisely and universally to define the term for several reasons. Firstly stories are teleological, the same content capable of hosting differing interpretations according to the teller's and listeners' purposes: hence definitions of history stories vary too. Secondly, standardized interpretations of stories cannot be made public since nuances of meaning are personally and privately constructed; the outlines of their meaning may be communicable but the fine details remain stubbornly individual, especially with younger learners. Thirdly, writers loosely interchange story and narrative and meanings vary with context. For instance the Shorter Oxford English dictionary (1983, p.2140) defined story in seven broad ways including an 'historical narrative or anecdote' a 'recital of events that have or are alleged to have happened' and a 'euphemism for a lie'. Pragmatically story parcels meanings and ascribes endings along such multiple lines, varying with its history, location and form until the synonyms hybridize: tale, narrative, myth, history, fiction, fact, faction. Just as shelter can be conceptualized as a universal need locally met by, for instance, Roman villas and Victorian terraces, so story's variants can be seen as particular responses to perennial cultural needs. Our mutual interest lies less in universal literary definitions, more in how people-as-teachers over time have used things called *story* for educational ends: and here we reconnect with Appleby et al.'s 'reformulation of old truths' (1994, p.284). Plato defined Ancient Greek educational story by its purposes. Since children's early opinions were 'usually difficult to eradicate or change . . . we should therefore surely regard it as of the utmost importance that the first stories they hear shall aim at encouraging the highest excellence of character' (Lee, 1974, p.133). Likewise a modern writer justified educational stories in moral, even heroic terms:

- building character;
- feeding the mind and the emotions;
- linking us to life;
- linking us to our past;
- helping adults and children to communicate;
- helping children develop skills;
- inspiring to greatness. (Shelley, 1990, pp.9–18)

The meanings and content of stories may be uniquely constructed, but elements of form traverse cultures: openings, endings, twists, tensions, plots and surprises enacted by characters. Commonly we 'speak of a life story to characterize the interval between birth and death . . . it is a commonplace' (Ricoeur, 1991, p.20). In educational terms this book collects classroom and staffroom evidence to test the theory that 'we never cease to reinterpret the narrative identity that constitutes us, in the light of the narratives proposed to us by our culture' (ibid, p.32). History as stories and traditional tales are two of these important narratives, not so much transmitted from generation to generation, but offered up for interpretation — 'proposed to us by our culture'. Homer's tales can never mean to us what they meant to the infantile or philosophical Plato, and modern constructivist theory holds that tellers and listeners define story by creating their own meanings:

> Story . . . keeps audience firmly in mind, for what counts as a meaningful story, or a good story depends on the listener who plays an active role in making sense of the story; it is the complicity of the listener which allows the story to repel the threat of meaninglessness. (Elbaz, 1991, pp.5–6)

Elbaz's *making of meaning* and Plato's or Shelley's *moral purposes* are educationally useful and ubiquitous. Put another way: 'Telling a good story at the right time is a hallmark of intelligence. One right time is when you are asked a question. Another right time is when someone says something to you and you respond with a relevant story' (Schank, 1990, pp.112–13).

Stories parcel meanings, make morals and mirror life: but 'educational stories' especially demand children thinking and doing as well as listening. Consequently, most of the storied teaching methods in this book prompt questions and elicit responses by slowing, speeding, repeating or freezing representations of narratives for educational ends. The ambition is to enable children to value but also interrogate narratives and thereby develop their own. This combination of narrative with enquiry defines what I mean by storied pedagogy. Stories are powerful in their own right but perhaps become truly educational when made susceptible to questions. A colleague illuminated this relationship autobiographically:

> I was an avid reader and always loved historical stories, so much of my knowledge and interest came via this route. My over-riding memory of history at Grammar school is of note taking . . . although the subject came briefly to life when . . . we had a lively male American teacher who *made us ask questions and told us funny stories* . . . (Marion Aust, September 1994, my italics)

In other words, the learner was drawn into the subject through lively narrative and questioning dialogue: they entered the world of the educational storyteller.

Journal — April 1993

How different are historians, storytellers and teachers? . . . The real difference may lie . . . in the materials with which they work . . . the information on which they base their stories. For novelists it is generally unverifiable in that it 'springs' from their imagination. In contrast, history demands attributable evidence, so that others may follow its process. As teachers, we need both.

This is important because for learners, many things seem fictional: like stories, realities are taken on trust from authority, the storyteller. Parents at first (don't touch that fire it will burn you) and then teachers (William the Conqueror was a Norman). These storytellers often offer their interpretations of reality *as* reality, and children test them out by experience. These experiences modify the story, and the children start to author their own existence. Often these modifications and communications occur through talk . . .

As children author more of their own stories, so they need access to others' as navigational aids. History, involving the reconstruction in our imagination of the past, is a mainstay of such growth. How can we enable that reconstruction to occur when many of the actors are dead or inaccessible? Sometimes, but in my experience not often enough, we recreate events then ask learners to enter them: to watch history as audience, enact history as imagined character, interpret history as author. This *entering* is a significant act . . . it relies upon the use of historical imagination fuelled, informed and bounded by evidence. But it is the nature of the evidence which differentiates the process from the authoring of fiction, not the process itself. To imagine life as a story, even if we do not believe it, is to empower the thinker as actor, author or both: to stimulate perceptions that other endings are possible . . . If teachers use methods which encourage learners to behave as if they were acting in an evidence-based fiction, might effective learning occur more easily?

Such views reflect more than my idiosyncratic experience. For Sarbin (1986) as for Collingwood (1946) both the novelist and the historian are narrativists: but as teachers' stories are live and interactive, 'In teaching, what is needed is a chance for children to give vent to their misapprehensions . . . This means using every possible means to "re-create" the situation, and talking through the agents' "problem" with pupils, both in considerable detail' (Lee, 1978, p.82). Contexts do not have to be storied to achieve this. All I suggest is that storied lessons and activities have helped me and others to motivate pupils into animated, focused discussion of particular historical problems; and provide many opportunities for teachers to monitor, channel or augment pupil knowledge.

Can Definitions from Practice Help Define History Stories?

To demonstrate these ideas in more concrete ways, I have abstracted evidence from a Y3/4 NC history lesson about Roman Britain. This might be summarized as a factionally storied, discursive, dramatic, whole class and small group role-play. It was led by an LEA drama advisory teacher (Sandra Redsell) at Redgrave School, a small rural 5–11 primary school. Extracts here point to two interesting mixes. Firstly, talk or story did not straightforwardly decentralize power: the *teachers* did much of the talking, creating frameworks within which precise, problem-solving pupil-led talk followed. What looked like a 'pupil-centred lesson' of role-play or discussion was actually teacher-reliant. Secondly, fact and fiction — or rather information and imagination — were blended. These extracts suggest that story can be informative, analytic and teacher-interventionist, in typical factual mode as well as imaginative, exciting and open-ended in traditionally fictional mode.

At the lesson start, the children were assigned roles as audience and actors, charged with the ultimate task of deciding how the story will end:

Sandra: We are the governors of Rome. And we are soon going to have a deputation from Roman soldiers who have been to Britain to try to explain why we should put an awful lot of money into sending an army of occupation to settle.

The children were allocated spoken, storied tasks with the 'governors' and 'soldiers' given discussion time to think their way into role. Then after nearly an hour of talk framed by the story, children authored an ending:

> *Sandra*: Alright a vote . . . those who think that we ought to go to Britain and settle there for all the excellent reasons we have heard? Thirteen. Those who think it is not such a good idea? Nine.

Yet the children were clearly and decisively brought in and out of role, demarcating 'their' fictional ending with what 'really' happened in fact:

> *Sandra*: Overall we have decided to go and overall that is what the Romans did. But I think that that point which our Senator over there made was a very good one, because what did eventually happen?
> *Child*: Some people around the country mutinied and they had to go back?
> *Sandra*: Yes . . . And that's when the next lot of invaders to Britain came which was who?
> *Pupils*: The Saxons.
> *Sandra*: OK, Well done. Line up at the door.

As well as specific sections of whole-class teacher-directed question and answer, unsubtle attempts were made during small group work to steer children from imaginative errors back to real information. Here the author was desperately closing avenues:

> *Grant*: When you went to fight in Britain, what were they like?
> *Child 4*: They were clever.
> *Child 5*: Yes, they had things to build, they had straw, which Rome doesn't usually have because Rome's a city and cities don't really have straw.
> *Child 3*: Yes, food for the cattle.
> *Grant*: What about . . .
> *Child 2*: They have, they have different clothes.
> *Grant*: They have different clothes?
> *Child 1*: There's different clothes there . . .
> *Child 2*: We have old clothes and they have silk.
> *Child 8*: And you could learn a different language.
> *Child 7*: Did they, what in Roman times?
> *Child 2*: Yeah.
> *Grant*: Do you think we could bring any of the people back? Could we bring them back to Rome?
> *Child 8*: Slaves!
> *Grant*: What a good idea!
> *All*: YEAH!

There were plenty of Lee's chances to 'vent misapprehensions' here (1978, p.82) but I made the classic teacher error of ignoring them. Then a 'right' answer ('slaves') was extravagantly celebrated. I was single-minded in pursuing perceived historical truth and prepared to manipulate the conversation to sate my insecurity. In consolation the children were even more ruthless:

Child 1: Because they um had different machinery um to make our clothes.
Sandra: That was . . . sorry?
Child 1: They have machinery, like they had machinery to . . .
[sceptical babble ensued with shouts of 'machinery'?]
Sandra: They don't know what you're talking about . . . No, don't understand.

These children were young but they knew that 'machinery' was anachronistically unacceptable. Within this storied mix, analytical historical reasoning and thinking were emphasized:

Sandra: We congratulate you on your heroic deeds . . . but are troubled by rumours . . . that we should not only invade and take things from other countries but that we must go and stay . . . we'd like to hear reasons.
Child 8: You get slaves and bring them back.
Sandra: Slaves, you can get slaves there?
Child 8: And then you can make money . . .
Sandra: How do you do that?
Child 8: By selling the slaves.
Sandra: We could do with a few more slaves . . . we've got a lot of building work.
Child 7: And you'd be a very rich country . . .
Senator: But who would want to sell these things and who would want to buy them?
Sandra: One question at a time . . .
Child 4: Other people in other countries will need their houses building as well.
Sandra: So we might get some slaves out of it? Maybe. What other reasons did we have then are you suggesting?
Child 5: Well there's iron so we can make swords, shields . . .
Sandra: There's iron? Is there? Ah . . . we do need continuous supplies . . . which will be jolly useful to us as we always need weapons. Alright, any other suggestions?

Despite the historically fanciful construct of soldiers persuading and senators voting this teacher language was purposeful, skilfully mixing fictional with managerial, historical and pedagogic demands. Storied roles were assigned but *controlled*:

Sandra: They don't know what you're talking about . . . No, don't understand . . . One more thing . . . Listen Senators please.

This control encouraged *historically accurate responses*:

Teacher: Remember why we are talking about it . . . we've got to have a reason . . . for spending lots and lots of money to let us settle in Britain.

Historical information was *communicated*:

Sandra: Listen . . . We've been told that it's got things like iron and silver . . . slaves . . . we reckon we can beat their army . . . it's got the sort of climate where we could actually grow things so it's possible to live there.

Questions stimulated thought:

> *Sandra*: What's your worry then if that happens? ... Alright, any other suggestions?

Since I am arguing that powerful teaching in history is characterized by this last question: *any other suggestions?* I now model it myself.

Suggestion 1: Might the Children's Responses Have Been Parroted Attempts to Please Teacher?

The children responded within a closely and carefully constructed framework: the lesson was planned and it showed. This did not necessarily inhibit real and independent learning. The controlling organizational and linguistic framework gave children the freedom safely to experiment with developing interpretations of history. The ten labelled children's utterances below show this:

> *Grant*: Who's got a reason?
> *Child 1*: Because there's nice land for the cattle. **[a]**
> *Grant*: It's a long way to go just for land ... I mean there's lots of land in between. It might be though, think about it ...
> *Child 2*: Because the gods told us to. **[b]**
> *Grant*: OK, you got a message from the gods that you ought to go there, alright ...
> *Child 3*: But that's lying ... **[c]**
> *Grant*: Well they might not be, perhaps the gods have told her.
> *Child 4*: Because the British might want to attack us, so we attack them ... because then the British will take over all the other countries that we own, won't they though. **[d]**
> *Grant*: They might ...
> *Child 4*: Um, it's a problem. **[e]**
> *Grant*: It's possible isn't it? Is there anything in Britain that might be useful to the people in Rome? That you might find there?
> *Child 5*: Well, you might find clay and ray and them that you might not find in other places and rocks and stone ... **[f]**
> *Grant*: OK, so one of those I think might be very important ...
> *Child 4*: Iron. **[g]**
> *Grant*: Yes, I think definitely.
> *Child 5*: 'Cos of swords. **[h]**
> *Grant*: Because it's quite difficult ... you know there's not many places where there's iron, certainly they must have needed that so that's a good one.
> *Child 6*: Different weapons. **[i]**
> *Grant*: In what way?
> *Child 6*: Like in Rome they all have swords and knives but in our country we have um, guns. **[j]**

Four of the children's utterances offered reasons starting with 'because' **[a,b,d,h]**. Three more similarly offered explanations **[g,i]** or speculations **[f]**. Another was an explanation in error through anachronism, subsequently admonished by the group **[j]**. The last two consisted of one question-cum-accusation **[c]** and an admission of puzzlement or failure **[e]**. Teacher talk framed these responses but they were thoughtfully learning about the causes of Roman invasions, how to think about thinking, discuss problems and verbalize solutions: these children were acquiring history as active and independent knowledge.

Suggestion 2: Was this Evidential History?

Was this educational story improper history? I was troubled by the lack of primary evidence and some talk within the framework was not accurately historical. Such observations influenced my development of teaching principles for history stories (see Chapter 6). But the fictional elements were preceded by a factually descriptive overview of the Roman empire. Together they created a factional history story into which children were invited as actors and authors. Within this they were then free to develop properly historical analytical and reasoning skills. Particular parts of this lesson may have been academically unhistorical (see Chapter 8) but the pedagogic framework enabling them to occur was 'let's pretend within the bounds of history'. Sandra's children achieved learning of a high quality because the history was embedded in clearly-explained, human-sense roles (Donaldson, 1978) and an accessible storyline, rather than prompted by an abstracted task such as 'discuss the reasons for the Roman invasions'. Using story, children were enabled to make historical knowledge mean something to themselves.

Can Stories Make Historical Imagination Reasonable?

Stories enable children imaginatively to enter a social world shaped by values different to their own as with the 'Vikings or the Greeks, because they have their own cosmology and Gods and heroes' (Hugh Lupton, storyteller, March 1990). The art of storytelling also manipulates time: 'not so much a way of reflecting on time as a way of taking it for granted' (Ricoeur, 1981, p.170). Myths, the distant cousins of story, may even transcend time:

> What is involved is not a commemoration of mythical events but a reiteration of them, a doing again of what was done 'once upon a time'. The protagonists of the myth are made present. One becomes their contemporary. This also implies that one is no longer living in chronological time but in primordial time, the time when the event first took place. (Polanyi and Prosch, 1975, p.123)

What do children make of these three axes: the manipulated storytime of Ricoeur, the primordial mythtime of Polanyi, and their own experiences of mainstream chronology in lessons, during which myths and stories are presented as historical evidence? The following classroom instance combined these elements. View this as another link in the chain of evidence connecting what theoreticians say about story with what I have witnessed or initiated in classrooms.

Classroom Evidence — Essex University

In 1992 a colleague (Derek Merrill, who has helped shape this account) and I taught several KS2 classes about Aztecs and European exploration (DES, 1991, p.29) on campus at Essex University, using a display of world maps in the university art gallery. My colleague designed the gallery teaching sessions, using 30 replica and original world maps dating from Ptolemy. I led the other 80 minutes in a common room. During this I showed five slides of Tenochtitlan, told the Aztec 'founding of Tenochtitlan' myth, told a story of why human sacrifices were needed to fuel the sun, emphasized the unpredictability of Mexico's climate and geology, organized group discussions of 10 photographs of Aztec artefacts and then set up a dramatic simulation of the market place at Tenochtitlan using translated sixteenth century evidence from Bernal Diaz. All children experienced both lessons, which were designed to show that Aztecs had a capital city as advanced as European equivalents and to imagine reasons for practising human sacrifice. One teacher followed this up with her 8- and 9-year-olds by asking them to write poetry. Two representative examples are cited below, perhaps simultaneously achieving sophisticated ends as:

- historically accurate;
- derived from analysis of a range of evidence;
- balancing interpretations of Aztec culture;
- representing evocative and descriptive historical writing;
- inhabiting the 'mythic' as well as 'chronologic' and 'temporal' planes.

Poem 1

Tenochtitlan the city of. . . .
nice smells,
of gardens
Tenochtitlan the city of. . . .
rotting smell
of sacrifice victims blood
Tenochtitlan the city of. . . .
terrifying sounds
of the unlucky people

Tenochtitlan the city of. . . .
selling sounds
of the massive market
Tenochtitlan the city of. . . .
little canoes
gliding down the rivers
Tenochtitlan the city of. . . .
sacrifices
the person life is at the end
Tenochtitlan the city of. . . .
carvings
of their scary gods
Tenochtitlan the city of. . . .
skulls
decaying on the rack

Poem 2

It was a place of greenery and lush vegetation
A place of beautiful stone carvings and great temples and jewels.
It was also a place of war trade and religion
A place of aggression war and captive slaves
A place of religion gods, and sacrifice and death
Carvings of gods looming over houses and temples
At the market people talking running screaming in the beautiful centre square.
An awful smell piping from the rotting skulls on the skull rack
And the nice smell coming from the flowers in the gardens
Tenochtitlan was a beautiful place indeed

I am not claiming that these children entered mythic time and gained direct access to Aztec culture, but I am suggesting that experiencing different evidence forms majoring on story helped them write vividly and historically about Tenochtitlan in present (Poem 1) and past tenses (Poem 2). Is such writing history? It seems more descriptive than analytic, it does not conform easily with 'enquiry questions' (e.g. Cooper, 1992, pp.6–7), and can lapse into literary ramblings from imagination, not evidence. Perhaps . . . but read them again with the following historical questions in mind:

What was Tenochtitlan like?
(1) 'the city of little canoes gliding down rivers'
(2) 'it was a place of greenery and lush vegetation a place of beautiful stone carvings great temples and jewels'

Were Aztecs unreasonably cruel?
(1) 'the city of carvings of their scary gods'
(2) 'a place of religion gods, and sacrifice and death carvings of gods looming over houses and temples'

Were Aztecs socially advanced?
(1) 'the city of selling sounds of the massive market'
(2) 'a place of war, trade and religion, a place of aggression war and captive slaves'

On what was their city based?
(1) 'the city of nice smells, of gardens, of rotting smell of sacrifice victims blood'
(2) 'at the market people talking running screaming in the beautiful centre square an awful smell piping from the rotting skulls on the skull rack'

To borrow from Collingwood (1939, p.39) *these were the questions to which our lessons were the answer*, though present only tacitly at the planning and teaching stages. They also lead to a further question:

Might myths and stories contain concentrations of cultural power into which children can tap?

Tellers' stories and listeners' stories — teachers' lessons and children's lessons — are not the same thing: 'knowledge does not enter the mind of the learner in the form transmitted ... never a straightforward copy, but a new, personal reconstruction' (Wells, 1992, p.286). Stories need translators (see Chapter 8). Equally stories and myths, like that of the sacrificial Aztec Sun God, exercise cross-cultural power by addressing common human problems. How otherwise can we explain the oral, the literary or *any* cultural tradition? If really ambitious of ridicule, I might extrapolate that some stories exploit humans as hosts. I would deploy the metaphor of myths as powerful viruses and argue that absorbing and controlling them should lie at the heart, not the edge of education:

> Stories think for themselves, once we know them. They not only attract and light up everything relevant in our own experience, they are also in a continual private meditation, as it were, on their own implications. They are little factories of understanding.

Since to make such a statement might endanger an educational historian, I leave it to a poet laureate to think the unthinkable:

> If the story is learnt well, so that all its parts can be seen at a glance, as if we looked through a window into it, then that story has become like the complicated hinterland of a single word. It has become a word. Any fragment of the story serves as the 'word' by which the whole story's electrical circuit is switched into consciousness ... take the story of Christ ... the nativity ... the crucifixion ... A single word of reference is enough — just as you need to touch a power line with only the tip of your finger. The story itself is an acquisition, a kind of wealth ... (Hughes, 1977)

Consider a history curriculum's *content*: lists of phrases and concepts to be taught (e.g. KS2's 'settlement, invasion, slavery', DfEE, 1995a, p.5). These appear not only dull, but isolated from meaning by being decontextualized. Now shine Hughes' light or my own on them, reflected from the two recently-cited lessons on Roman Britain and the Aztec empire. View *invasion* not as a technical term but as a 'word by which the whole story's electrical circuit is switched into consciousness, and all its light and power brought to bear.' Historical content, concepts and vocabulary suddenly become clear when I conceive of them like this. When words live as stories, anything can happen.

Consider a history curriculum's skills (e.g. Key Elements, DfEE, 1995a). I previously raised the possibility that some children's poems answered more demanding historical questions than were explicitly asked. Hughes' idea that 'stories think for themselves' may provide a partial explanation. Might stories teach historical thinking skills as well as content, naturally and organically? Jacquie Whiting taught KS1 children when I interviewed her, often through storied drama:

> You pick up things and you don't even realise . . . We'd done this work on the Orford Merman and talked about different versions . . . Sometime later on . . . when I was doing some RE . . . one little boy said that's the same isn't it, different people telling different things? (Interview, March 1993)

As I read them stories containing in familiar and time-tested packages the trinity of knowledge as content, information and understanding. If made well, told clearly and listened to with mind as well as ear, stories may linger to be learnt from when ready. Given excellent telling, story-listeners may even receive a lifelong gift to revisit at progressively higher levels of understanding. As my colleague said of the Aztec teaching just cited, 'The combination of evidence was crucial but the story will probably be remembered longest . . . story was the starting point . . . story captured the imagination' (Derek Merrill, 1996).

Teaching history is therefore a humanitarian venture. 'The teacher's job is to make things make human sense so far as the task itself is concerned' (Lee, 1984, p.111). Although stories seem to assist this, could what I am discussing be called 'drama'?

> Story has often proved difficult to handle as it presents a completed narrative, whereas drama is concerned with the building of a narrative. Furthermore, story often places a teacher in a position of dominance and supremacy which inhibits children's ability to take charge themselves. (Fines and Verrier, 1974, p.23)

Persuasive grounded evidence has illustrated a range of drama-led history teaching strategies (e.g. Goalen and Hendy, 1994), and many 'teaching ideas' in this book are drama-based. Without wishing to detract from drama in any way I prefer 'story' as an umbrella term. Drama traditionally intimidates secondary school history teachers, remaining a 'very occasional, albeit spectacular, weapon in the teacher's armoury' (Shemilt, 1984, p.67; see also Culpin, 1984; Goalen and Hendy,

1994; May and Williams, 1987). Drama raises arguments in its own right (e.g. Abbs, 1994, pp.117–22) and became entangled with historical *empathy*, now a devalued word (Knight, 1989). Story may also coin broader educational and cultural currencies, covering the participative and imaginative activities of classroom drama but being recognized more widely than they. Crucially for pedagogues, story explores how (generally) single authors can make the complex simultaneously simple, accurate and engaging. Teachers have to do the same:

> A common assumption is that simplifying entails removing narrative or contextual detail . . . In fact if an event is complex . . . nothing can be done to alter that . . . Teachers are not mistaken in their intentions when they set out to simplify . . . but are mistaken if they think that conciseness is the same as simplicity. (Shawyer, Booth and Brown, 1988, p.216, my emphasis)

The answer is not to skate over or slenderize content, making history hard to enter or dispute (Hull, 1986, p.92). Rather, to be counted as educational the stories we call history should be interrogated in depth. This takes time and, as the next chapter argues, it takes talk:

> Of course the past must be simplified. Of course children will need some stories to make sense of it. What they do not need is the story of the whole British past: they need a workable framework . . . room for manoeuvre . . . Pupils need time to examine passages of the past in detail . . . Children need time to talk. Teachers need time to listen. (Lee, 1991, p.62)

In Practice (3)

Teaching Through Written History Stories

The following teaching ideas apply to written texts and aim to develop critical literacy. They can be used across age ranges but are angled here towards supporting younger children.

1. Beginning to End

Children are asked to re-order a set of pictures from a storybook, or a thematic collection of visual sources to tell their *own*, the *'original'* or *different versions of the story*; and always to explain why and how they have done so. Variations include grouping pictures to show the beginning/middle/end, comparing versions across classes, and distancing pictures appropriately to represent different timescales, relationships or endings.

---

2. Gaps and Endings

A set of pictures and/or words tells a real story from the past — but some are missing. Children speculate about the gaps by talking, drawing missing pictures or writing key words. It is particularly powerful to speculate about endings. Through this children gain awareness of the range of possibilities open to people at the time: but also that history stories define endings by describing 'what actually happened'.

3. Who's the Heroine/Villain?

A set of pictures and/or words describes people from an historical story. Children discuss characters at the centre or edge of the action, placing their pictures appropriately on a drawn circle. If you wish to be more sophisticated and discuss how books came about — introduce pictures for the author, illustrator or publisher. If you want to discuss historical evidence in particular, talk about how sources mentioned in the book came to be there (e.g. a reporter talking on television, somebody writing a letter).

4. Why Did They Do That?

Ask children to devise questions of particular pictures or extracts from a story, concentrating on the reasons for and results of people's actions. A variant is for children to have a larger set of pictures and *pair them off*: this picture happened because of that one . . . and so on.

5. What Changed or Stayed the Same?

Using pictures or discussion after a story, ask children to pick out things that changed/stayed the same over the timespan of the story. Pictures can be grouped or cut up under these headings, or words grouped and written for display.

6. Is It Like That Now?

Comparable to the above, but asks children directly to compare the past of the story with the present day. What is the same or different about houses, food or school? An extension is to use different stories about the same themes to compare two different historical times and the present.

7. Titles and Captions

Children devise main, chapter or section titles for an historical story or information book having been given most of its pictures and/or words. Captions works similarly but focuses upon naming particular words, pictures, objects or evidence in the text. Either can be compared to the original, giving multiple opportunities for examining books in general and using historical words in particular.

8. Editing

Simply, children are asked to reduce the number of pictures but still tell the same story. A more complex version edits words too. Which ones would children choose? How does the story alter? To increase support, alternatives can be offered from which children choose.

9. Real or Made Up?

Children are given sets of historical pictures, characters, stories or books about a time they have been studying. They discuss and divide the pile into stories that are fact and fiction — real or made-up. More sophisticatedly, can they explain their choices, or place materials on a graded continuum?

10. Versions

Find a well-known story or event in different books, texts, pictures or videos (e.g. Fire of London, World War Two evacuation). How many different ways can children find of comparing these differ-ent versions? (e.g. length, pictures, types of word, when they were written, trustworthiness). Can the children find any different evidence or ideas in them?

11. Spin-Offs

After reading a story about school 40 years ago, children collect stories from grandparents and turn them into a book. After read-ing a story about an old house or what's under the ground, chil-dren explore historical houses, research archaeology and convey their understanding through painted or dramatic displays. Stories here are launch pads for the children's own imitations or versions of processes first encountered in the stories of others.

12. Rewrites

After telling, children discuss how they could retell the same story in another form (e.g. as a poem, song, drama, storyboard for a video). To keep this specifically historical, give children from different groups three common pieces of 'evidence' (e.g. an object, picture, place or word) which have to be in the new version. Even though they generally want to, children do not have to rewrite the whole: simply discussing ideas may serve your purpose.

13. Key Points

Having listened to an historical story, ask children to pick out the most important: person, thing, event, place or meaning. More sophisticatedly, can they tell the story again from memory and, using an audience or a tape recorder, discuss what they left out or changed from the original?

14. Talking Bubbles

Pictures of story characters (or authors) are reproduced with speech bubbles attached. After hearing the story, children devise questions and answers and write or have them scribed in the bubbles. Variations are to have characters 'talking' to each other. To make this specifically historical, the teacher models historical words or concepts beforehand.

15. Let's Play . . .

If young children have watched, heard or read an historical story they will often spontaneously want to 'play' it. This play may be harnessed by teachers, so long as their adult values are not allowed to spoil or dominate what is happening. Cooper (1995a, pp.63–76) gives useful and detailed examples of how play can support historical learning.

16. Pot Luck

The names of the characters in a story are written down for children to draw out of a hat. When the story is told or retold, the children pay particular attention to 'their' character. Afterwards they may be asked to explain why they did a particular thing or to say more about that character through words or actions.

17. What Might the Moral Be?

Most children have experienced fables, such as Aesop's and it can be useful for children to discuss the possible moral of an historical story, compare individual interpretations or introduce other and comparable stories. I believe this can be achieved without presenting history stories as 'having' a moral intrinsic in their events — as in religious or mythic stories. Do you agree?

Chapter 4

History, Stories and Talk

I know quite well why I became an historian. Quasi-historian, as one of my enemies put it . . . It was because dissension was frowned upon when I was a child . . . Argument, of course, is the whole point of history . . . (Lively, 1987, p.14, 'Claudia' talking)

Why Should We Value Educational Talk?

Research into oracy helped fuel an educational revolution in the 1960s: 'the learner is active; he must co-operate in his own education' (Wilkinson, 1965, p.59). Interest in talk as a necessary educational power broker and breaker continued into the 1970s when for the first time 'oral communication and exchange . . . became a separate field of enquiry' (Dixon, 1988, p.24). Classroom talk about history figured in this burst of activity (e.g. Barnes, 1976; Hull, 1986; Rosen and Rosen, 1973; Tough, 1979; Wells, 1986). Such scholarship inspired my own teaching and underpins curiosity about the following classroom behaviours.

Talk as Thinking Out Loud

It is difficult to be clear about the personal chronology of important ideas. Do we see them in practice in classrooms and *then* discover them in books, or see them in practice in classrooms *because* we have discovered them in books? Barnes has illuminated my vision since reading him during my PGCE year of 1981–82: 'Speech, whilst not identified with thought, provides a means of reflecting upon thought processes, and controlling them. Language allows one to consider not only what one knows but how one knows it' (1976, p.98). Since then, teaching children to communicate has fuelled my pedagogy. The publishing bonanza offered by NC history (Cooper, 1995b, p.vii) contributes much: a plethora of books, videos, packs, posters and CD ROMS now communicate history's content and ideas, in multiple forms and genres. Yet teachers often do not use history textbooks effectively (Ofsted, 1993, p.36) and typically 'circumvent the difficulties of (history) textbooks by offering their own explanations' (Wishart, 1986, p.148). Teachers' and children's talk remain crucial for mediating such resources, and written curricula and textbooks seem helpful only so far as they engage us 'through speech with important aspects of the social and physical world' (Barnes, 1988, p.48). Smith similarly underscored purposeful, contextual language learning (e.g. 1985). Setting himself against the 'prevailing view in education' that 'learning is work' he promulgated:

An alternative informal view, at least 2000 years old, and well known in every culture of the world. It is that learning is continuous, spontaneous and effortless. It requires no particular attention, conscious motivation, or specific reinforcement and is not subject to forgetting. The learning is social rather than solitary. It can be summarised in seven words: We learn from the company we keep. The official and informal views of learning are diametrically opposed . . . (Smith, 1991)

Such informal, cooperative ideas about educational talk have an undistinguished history in our modern educational system even though 'the majority of teachers accept or pay lipservice to them' (Jones, 1988, p.28). 1980s teachers often maintained classroom control through classroom talk with curricular divisions being 'too sharply and hierarchically defined for pupils to experience many opportunities to enquire, experiment, or create their own meanings' (Stubbs, 1983, p.35). Such deeply ingrained difficulties were ignored by a new 1990s orthodoxy which claimed to see 'too little direct teaching . . . the classrooms where children make most progress are . . . where the bulk of the lesson is taken up by the teacher explaining, questioning, pushing back the frontiers of the children's knowledge: in short, by teaching' (Ofsted, 1997, p.6). Storied structures and methods can support both ambitions: teacher-talk *demonstrating* how to understand and pupil-talk *developing* it.

Talk as Analytic Anecdotalism

Students' anecdotes are traditionally seen by teachers as disruptive (Jackson, 1983) or irrelevant in history lessons (Hull, 1986), and anecdotal thinking is typically marginalized by schools in favour of styles emphasizing 'definition, abstraction, conceptual analysis, and rigorous canons of evidence or proof' (Robinson and Hawpe, 1986, p.123). The discussion-led 1970s HCP (Gardner, 1993) was initially suspicious of anecdote's role, only changing direction after finding that for students with low self-esteem, exchanging anecdotes provided 'a way of approaching discussion without placing their ideas at risk of ridicule from their peers' (Elliott, 1991, p.22). Anecdote offers a way for individuals to feel their way into a controversy, of 'making abstract argument more concrete' (Berrill, 1988, p.66). Anecdote as oral history remains an historical source independently accessible to young learners and, as personal storytelling, an especially suitable enquiry tool (Cramer, 1993). Telling a story also tests the teller's persuasiveness and authenticity for if it is not well-told the story fails and 'the story maker must either forsake his views or alter them to accommodate the views of others' (Robinson and Hawpe, 1986, p.117).

Talk Rendering Historical Knowledge as Negotiable

Lee's extensive research into pupils' historical thinking showed that at classroom level children's understandings were still unpredictable: 'This necessitates talk with pupils in an open framework which allows children to develop their ideas and teachers to monitor them' (Lee, 1991, p.58). Talk in particular leads to unexpected achievements. As a teacher commented after a spoken and dramatic lesson 'when you present the evidence in a story, you dramatise it . . . they appreciate the kind of things that if you gave them a written question would throw them completely' (Ipswich Y3/4 Teacher, November 1991). Concentrated and purposeful talk entails hard thinking and can 'force progressive accommodations to new data' by pupils (Shemilt, 1984, p.79). Two-way classroom talk also democratizes cognition: 'If the teacher were to teach, as it were, himself, he would run the risk of imprisoning the pupil in his ideas' (Stenhouse, 1967, pp.137–8). Elliott argued that dialogues between teachers and students were vital 'procedural principles governing any educationally worthwhile induction into knowledge' (1991, p.142). Equitably distributed, talk devolves educational power by giving children passports to enter 'the discourse of the discipline and take over the ways of thinking . . . made public in that discourse' (Wells, 1992, pp.294–5). This entails serious and active listening by teachers to pupils, and vice versa. For Polanyi history teaching relied upon 'the pupil's intelligent co-operation' (1983, p.6) while Shawyer and others similarly emphasized history teaching as listening: 'Enabling pupils to voice puzzlement is an important pedagogical skill' (Shawyer et al., 1988, p.218; also Curtis and Bardwell, 1994; Wilson, 1985). A massive weight of experience and theory points to teaching history as a cooperative discipline into which talk and story tempt learners to participate.

Talk Making Stories Thoughtful

Reading historical evidence and reading stories are related, almost 'subliminal' acts with 'unspecifiable' clues analysed intuitively. 'Such is the effort by which we enter into the intimate structure of a skill . . . the method by which an historian explores a historical personality' (Lavender, 1975, p.31). Children need such textual challenges. They are entitled to diverse, even puzzling historical experiences: to exploring curricula balanced between obvious processes such as phonics or dates, and the endless mysterious questions of history such as *how* and *why*? Likewise myths and legends are important for children to experience, more so even 'than to understand . . . talk will be one of their most vital responses' (ibid, pp.36–7). Such talk can simultaneously be *external and verbal* and *internal and thoughtful*, rather as Socratic dialogues resemble both conversations and thinking (e.g. Hamilton, 1971). If subjected to 'asking, listening and testing', might spoken history stories, like Socratic dialogues, form 'the ideal paradigm for developing critical intelligence'? (Abbs, 1993, p.3). Questions and stories are not necessarily different creatures. Storied verbal experiences also dig foundations for written skills, as children learn through talk internally to anticipate and impersonate 'audiences in advance . . . as when we compose narrative texts' (Herrnstein-Smith, 1981, p.230). Bethany, aged nearly 13, described this phenomenon when interviewed about her historical story writing:

Well you read it and you like think that you're back there talking to someone and you can just feel that it's wrong or not . . . sometimes I say what I am going to write and then write it at the same time just to make sure it's OK . . . I always stop after a bit and ask my friends what it sounds like. (Blackbourne Middle School, 1990)

Consensus almost exists on the importance of sustained, spoken enquiry: 'pupils responded particularly well when they had opportunities to discuss ideas . . . Their involvement was usually greater when they were able to speak or write at length' (Ofsted, 1993, p.9). I interpret this as history needing a storyline, and one that motivates learners to engage with it through talk. Skilful talk can 'transfer the enthusiasm of the teacher to the class more quickly than any other method' (Blyth, 1989, pp.75–6). Direct teaching supports children's linguistic development through the 'unlocking of word-hoards . . . History, with its emphasis on the understanding of human actions . . . affords particularly fruitful opportunities for such work' (Jones, 1988, pp.169, 171). Mixing talk and questions with story and meanings boasts deep educational roots. Two of Western Europe's most important teachers spoke much, from the east, but wrote nothing. Both continue to be cited in contemporary educational writing: Socrates typically being associated with critical dialogues, Jesus with contemplative parables (e.g. Abbs, 1993, p.1; Gudmundsdottir, 1991, p.209). Such traditions are also represented in classrooms, as the next section shows.

Oral History in Schools

I became interested in oral history in the early 1980s. It seemed to offer motivating, meaningful opportunities for children directly to analyse evidence at source, 'to question the makers of history face to face' (Perks, 1992, p.6). By the early 1990s my thinking and oral history's practice (Dunaway and Baum, 1996 introduction) was more complex. Was it oral history's narrative nature that made analysis and questions more accessible? Might storytelling's educational power derive from mimicking or recreating talk with the dead, making the past beyond living memory a questionable, oral history? (Yow, 1994). Oral traditions are multi-layered sources of evidence conveying 'not only the interpretation of the witnesses to an event but those of the minds who have transmitted it' (Vansina, 1980, p.276). A few extracts from empirical classroom work illustrates these themes:

Oral History with Children Aged 4–11

4 and 5 year old children are perfectly capable of reminiscing their way into a life story:

Grant: So, can you remember now about the first day that you started school?
Girl 1: I remember when I first started, well I was really scared of Mrs T . . . and when I was doing her work I was really scared to come and show her.

> *Girl 2*:   I went to my friend's and, and I saw Ian and, and he sat us down here then I becamed his friend . . . I made Ian's friend.
>
> *Boy 1*:   I was shy of Mrs T . . . (Suffolk Primary School, 1992)

Questioning prompted socially constructed memories (Boy 1 echoing Girl 1). Although such cooperative memorizing has many *educational* attractions (Hazareesingh, 1994) it has *historical* dangers: 'At best . . . the history not of people in general, but of persons in particular . . . at worst . . . little more than a misguided exercise in ancestor invention' (Cannadine, 1989, pp.191–2; Seldon and Papworth, 1983; Stenhouse, 1978) How do teachers balance such benefits and dangers? Hilton described mixing oral history and historical storytelling work with 8-year-olds, for which a special interviewee was needed:

> An older person . . . skilled enough to shape their own experiences in the twilight zone between speech and written story . . . The person needed to be able to tell his or her stories in a small way but with a sense of historical importance . . . I found a retired primary schoolteacher. (Hilton, 1989, p.188)

School teachers are not the only people capable of such a mix. In 1992 I listened to a class of Y1/Y2 children from Southwold interview Peter Gerrell, a local writer, radio reporter and oral historian who also happened to be blind. These factors combined to forge a narratory power so extraordinary that I cite it here. This sustained through talk alone and without eye contact the historical interest of an infant class for 80 minutes. Others have noted an oral history informant as 'a story-teller par excellence' in language 'simple and expressive — even colloquial in character' (Wilson, 1985, pp.71–4). Peter Gerrell's approach was similarly storied. Apparently polished through writing, his was a distinctive, subtly romanticized, even mythologic telling. Simultaneously it modelled how to question, as well as listen. For instance, I asked him to explain to the children how he 'found out about the way things used to be'. He cited one of his many hundreds of sources, 'each one' with a 'different story to tell'. Then he advised the children:

> Listen around you . . . I hear stories or I hear people speak . . . Last week I heard about an old horseman . . . I went and interviewed this old man, with a tape recorder like Mr. Bage is doing here now . . . I hear . . . I read . . . Mainly I learn by word of mouth . . .

After so discussing methods, Gerrell drew on the class's visit to a museum to stimulate questions, then moved on to those the children had prepared. As a teaching episode it was seamless: the prompting and answering of questions by telling stories. These stories synthesized his own and others' experiences. Peter Gerrell was 60 when he died, eight months after this lesson. *Historically* I still find it hard to decide whether its composite nature (e.g. Henige, 1982) made his evidence more or less trustworthy as testimony. But *pedagogically* I trusted it: the sentences were short, the language simple, the imagery and affective appeal captivating. In this typical extract three particularly pedagogic aspects are highlighted:

- *linking* his experiences with children's;
- *questioning and answering* sensitively;
- *highlighting change* and its causes and effects.

Extract from the Oral Story

Commentary

Child 1:   *What did the children play with, if they didn't have plastic things . . . ?*

Children's *questions* led the story.

Mr Gerrell: If you could only be taken back to do some of the things that those children did then, you would be very reluctant to leave them be. They were lovely games they were really. *The things we did in those days without the plastic things . . . Children did their own entertainment . . .* kite flying, *hopscotch — you've all played hopscotch of course —* conkers, tag and sparking hobnails. Now sparking hobnails you will never see done today, but *boys in those days wore hobnail boots . . . boots with studs in the bottom. Now why?*

Gerrell *linked* past and present by introducing historical alongside contemporary experiences and terms.
A *question* extended those

Child 2:   *To stop them slipping over?*

Mr Gerrell: *Yes, to stop them slipping over, and another thing was to make them wear — to keep them on boys' feet as long as they could.* When my mother went to school, she had seven brothers. They had one pair of hobnails between all of them . . . so for six days my mother walked to school — two miles . . . in bare feet, but on one of those days she was allowed to have the hobnails. They shared the hobnail boots out. Those hobnail boots were expected to last a child practically the whole time they were at school, because they were handed down from one to the next. *The games, there were so many games. Unlike today, we were allowed to go nearly everywhere. You children are not. It is too dangerous and it is a different world. It is a dangerous world today.* When I was a little boy there were woods all round here. There was a wood at the end of this road . . . and we could go in those woods. We used to climb trees . . . play tracking . . . go on the rivers fishing . . . *I'm sure you children would have loved every moment of it. It is a different world today; it is a dangerous*

links, stimulating interest in *change*. The child's *answer* was valued and added to.

Personal experience *linked past and present*, along with a value-laced judgment about *change*. In imaginative terms children momentarily inhabited a past world then were reminded this was impossible because of *change*.

> *world . . . in those days it was different.* It
> was a lot better place because, I think there
> was not the danger there is today. The   A possible cause
> games were many and they were lovely.   and effect of
> *You would have enjoyed them. I'm sure you*   *change* was
> *would. Plastic wasn't invented you see.*   recapped.

Naturally *historical* as well as *pedagogical* questions arose. Which aspects were based on direct personal experience? Did the narrative display a 'golden age' syndrome? Oral or storied evidence needs such scrutiny but KS1 or KS2 teachers especially need to balance methodological with educational anxieties. The class teacher here did so by asking further questions, diversifying sources and discussing stories and truth. Another interview with an experienced primary headteacher from the early days of NC history illuminated this need for teacher confidence and translation of the discipline.

> We will enjoy doing history but as for getting through the national curriculum . . .
> There is no way you can teach properly and teach that . . . I mean we had the
> Diocesan director of education on the phone one day and she said, 'what does this
> mean?' and she read me a piece of it. And I said, 'Well you can hear my staff
> laughing because they've just read that bit.' That was the history document, the
> day it came out . . . (Betty Woollams interview, 1990, see also Chapter 8)

Meanwhile some of her 8- and 9-year-old children were simultaneously winning a Suffolk history prize for evidence-led enquiries into the village mill. Unprompted, she described how their accumulated information was organized:

> I rationalised it into a story . . . into what you would do in a primary school . . . all
> the things we'd talked about and done . . . I got an enormous sheet of paper and
> laid it all neatly out . . . we discussed the colours and everything else . . . We put
> the story on one sheet, with the photographs . . . (ibid)

This project emphasized practical and spoken enquiry through letter-writing, interviews, fieldwork and documents, with reference to only one book of secondary information. Similar benefits from another oral history project on World War II evacuation, culminating in a community drama, were described by the headteacher of a Norfolk 5–11 school:

> What the children really got out of this was . . . how to find out about things . . . they
> went away and talked to their neighbours and grandparents . . . spontaneously want-
> ing to find out . . . a number of them have said it's the best project they've ever
> done . . . I think it was that it wasn't book based . . . that you didn't know what you
> were going to find out or who was going to get in touch made it quite exciting. (In
> Bage, 1995, pp.245–7)

These teachers valued oral history for enabling children directly to interrogate historical evidence — but also as powerful stories. Both teachers worked towards narrative outcomes (exhibition, drama) expressed in story forms. Peter Gerrell combined these spoken traditions: oral history led by enquiry in its questions and stories in its answers. Moving from question to answer, from enquiry to description, from past to present, from personal to general all while gripping infant attention. I felt in awe of him. He had no degree, teacher training or INSET: his skills came from an older pedagogy.

Lowestoft Journal, October 2 1992

Peter Gerrell, writer, raconteur and rural philosopher, died at the James Paget Hospital yesterday morning, aged 60 . . . He was an exceptional man . . . who could always see the best in people. Although he hankered in many ways for the tranquil ways of days gone by, he had all the time in the world for the youngsters of today. Peter Gerrell had little formal education, yet he had a way with words that won him hundreds of friends . . .

If as teachers we could learn to speak a little more through such well-rooted traditions and a little less through modernist and bureaucratized curricula, might we too 'win friends'?

More quantifiably, school oral history has been identified with 19 different benefits 'in varying forms, at any stage of social and intellectual development between the ages of 5 and 18' (Thompson, 1988, pp.166–9; also Perks and Thomson, 1998). From KS1 classroom experience it has been argued that oral history alongside activities like artefact handling are 'more likely to be remembered than any didactic approach to the teaching of history' (Peck, 1992, p.42). Was Gerrell teaching *didactically*? The Shorter Oxford English Dictionary defines this tricky word neutrally: 'Having the character or manner of a teacher; characterized by giving instruction; instructive, preceptive'. Agreeing with some of Woodhead's sentiments (Ofsted, 1997, p.6), teachers need to reclaim didacticism. As the Gerrell example showed, oral history can entail highly concentrated didactic experiences that nevertheless rely upon questions and appear educationally effective. I used to think, like Peck, that oral history decentralized power. Now I am arguing that in some oral history, adults do the length of talking which gained didacticism a bad name in the first place. Cooper epitomized such uncertainties: 'adult–child interaction is important if it is not used to transmit didactic information, but in order to help children to understand a question and how to answer it' (1995b, p.21). I am explicating a tradition of pedagogic storytelling and prompting which values and reconciles both: teaching as concerned with the answering and asking of questions and the prompting and telling of stories by definition. Oral history is not therefore 'a radical departure from conventional pedagogy' but rather a 'very old method of teaching' (Neuenschwader, 1976, p.38). Perhaps this is why, despite being identified with 'new history' (Purkis, 1980, p.2), it became embedded in KS1 and KS2 NC history. It also explains why positive views of questioning and enquiry in school oral history are practically consensual amongst educators (e.g. Andreetti, 1993; Aris,

1993; Claire, 1996; Cooper, 1995a, b; Cramer, 1993; Hewitt and Harris, 1992; Redfern, 1992).

Oral History with Students Aged 11–14

Motivation (Cleaver, 1985); bilingualism (Cramer, 1993); PSE (Orchard, 1992); inter-generational 'socializing' (e.g. Dodgson, 1984; Watkins, 1992) and inclusivity (Morris, 1992) have all been claimed as benefits of oral history in secondary education. Bluntly, students like its affective and interpretative nature: 'because as they talk you can sort of bring up what they really feel . . . the expression in their voice. Whereas written down you might read it wrongly and so you don't really get what they're feeling' (Bethany, aged 12, interview 1990). Mark (aged 12) expanded this, describing an information and also interest-rich interview with an elderly neighbour:

> His ship got hit once down at the end and some people died and he survived . . . He stumbled . . . I felt a bit guilty . . . It's better than just looking at the paper and writing it down because . . . It don't feel the same when you're reading something . . . Books don't say it in the same ways either . . . you actually get what they think and thought about the war . . . When you look in books and stuff you don't get so much information. (Interview, 1990)

This age group has more experience in handling, comparing and evaluating a range of different sources. Interviewing students across the ability range after a Y8 history project leading with oral history, Jennifer revealed that she 'liked the interviews. It's good hearing yourself on tape.' Ewan 'found out quite a lot from the interviews and the books. We didn't get that much out of the questionnaire . . . the interviews had more open questions so they told us more.' Claire's questionnaire on women's role revealed pitfalls: 'Some of our questions you thought were really good but when you like asked people, they didn't really understand what you were saying . . .' Oral historical sources were the most useful because 'In books they've got a certain amount of facts but in the questionnaire and the interview you found out exactly what people thought, whether they thought the same as what it said in the books.' Toni echoed this: 'I preferred the interview . . . because you found out a lot more and people could talk, they didn't just put yes and no.' Unexpectedly and unprompted, both girls then cited stories:

> *Toni*: What I found most information from was the interview. There was a question 'what things can you buy now in the supermarket that you couldn't then?' and I thought Mrs Evans was just going to say 'oh this that and the other'. *But she actually told a story* . . . there wasn't Fairy Liquid, like you had to mix up all this gooey stuff to clean the ovens . . .
> *Grant*: She told you a story?
> *Toni*: *Well she sort of told a story for each question*, like 'my brother he used to drive a taxi and blurb blurb' but that was really good. I liked that. I found out a lot of information, yeah.

Claire: *Like my mum would tell the stories and you could actually imagine it happening, like you had to really scrub out the oven . . .* she said that there weren't any televisions and I thought 'what did you do all day'? . . . People say things, you don't write them all down but you remember . . . you found out things but they weren't to do with the project.

Even books benefited from stories and oral history, as Richard explained:

I thought the books were better because it tells you more . . . like I done housework equipment for women and it tells you what the women said about the washing machine and everything . . . that was useful . . . one of them if their washing machine ever broke down, if they turned it on their side it just started again.

The spoken printed word seemed best . . .

In Practice (4)

Storied Teaching Methods Emphasizing Talk

1. Oral History

Children prepare and then interview respondents. Important stages are:

- practising communication skills;
- securing interviewees;
- deciding questions;
- choosing recording and reference methods;
- analysing the evidence collected;
- comparing evidence;
- drawing conclusions;
- framing further questions relating to other historical evidence;
- communicating results;
- thanking participants.

Memory-jogging starting points include: maps, walks, photographs, printed or spoken stories, buildings, scars, toys, food, clothes or key national events. A wide literature exists relating to this over-simplified summary.

2. Performing Stories

An historical story is amplified through research, reduced through artistry and then told using words, sounds, mime, expression, questions and evidence. Some teachers do not physically involve audiences. I prefer to, by stopping the historical action at key

---

points and including learners through mime, talk, decision-making, advising or acting. Performance can be adapted to all media: pictures, models, music, poems or puppets can represent stories from the past and be brought alive through talk.

3. Storyboxes

From a box or bag children describe artefacts or representations of historical evidence and characters (e.g. a Victorian penny, book, school photograph, reproduction toy) then devise a written or spoken story about them. Variations include:

- using models of artefacts or figures;
- asking children to make their own storyboxes;
- impromptu stories based on drawing objects singly out of a bag;
- giving children the same sets and comparing resultant stories;
- devising toy museums including layout, signs, audio-visual effects etc.

4. A Storied Production

The story of an object is told through the actions or materials needed to produce, transport, sell, use and preserve it. A witnessed example from Frances Sword (Fitzwilliam Museum, Cambridge) took the following stages in the life of a Greek vase, during which children questioned, demonstrated, mimed or listened to the story:

- digging clay by slaves;
- the potter and wheel;
- firing and fuel;
- transport-by-pack-animal;
- the agora;
- the purchaser and purchaser's house;
- the vase's final resting place, and why;
- archaeologist;
- museum conservators and visitors.

The author has successfully used the same principle for modern examples, e.g. Victorian bricks, World War II bombs.

5. Talking Objects

Similarly a real or imagined object is 'brought' to a circle. Its stages are described and explained individually or cooperatively. Questions are asked about the object's owners, life, purpose and experiences (e.g. Tudor witches bottle, Victorian sovereign).

- making;
- using;
- losing;
- leaving;
- finding;
- stealing or preserving.

6. The Mantle of the Expert

In this 'tasks are set up which allow (children) to practise and grow into experts' (Towler-Evans, 1997, p.106). Adopting roles such as archaeologist, museum manager, documentary maker, teacher, journalist, conservator or antique dealer. Children are asked to:

- Enact a specific job, e.g. 'You are an archaeologist. The phone rings with news of an exciting find. What do you pack in your bag?'
- Approach a piece or collection of evidence to solve an historical problem, e.g. 'How would you turn these pictures and letters into an interesting museum display?'
- Report on an interesting case, e.g. 'How did you write up an archaeological find for the local newspaper?'

7. Commentaries

Children devise and perform voice-overs for a historical film or set of visual historical images. This works particularly well for archive film or photographs. Commentaries can be taped to make audio guides or radio programmes about 'historical sites'. Such stringing together of disparate events or pieces of evidence into longer explanatory narratives resembles Applebee's (1978) model of stages of storied development.

8. Ceremonials

Using relevant evidence for a special occasion from the period in question (e.g. a building to open, queen to crown, memorial to commemorate, etc.) discuss: invitations, speeches, music, decoration, architecture, interpretations and views of the event. Mime or act out the event and its repercussions, allowing the dramatic action to be stopped, questioned and analysed.

9. Statues

Choose an historical character or event to commemorate. Divide the class into 'statues' and 'sculptors' with the latter arranging the former for two minutes. All statues are viewed and can be questioned by the 'sculptors'. Captions/plaques can be written. Children experience both roles. This focuses on interpretations of history and works best as a summative activity.

10. Role Play

Role play asks children to simulate involvement in a real or imagined historical community or event. It can result in large and detailed living history projects or simpler classroom mimes for participation and questioning. At either end of the scale it is useful to group children (e.g. as families or apprentices). This necessitates cooperative meaning-making. Stopping the action with 'events' (e.g. theft, invasion, trial, or arrival of a stranger) is another classic strategy to involve participants and pose problems requiring thought and discussion.

11. Any Questions?

Having been given evidence, children prepare to interview an historical character in a familiar format (e.g. news, chat show, magazine). Questions are prepared, ordered and compared. Children may answer questions, progress to being that person in-role or write up the interview.

12. Freeze Framing

After accessing evidence children construct a series of static mimed pictures to explain and represent the stories of an historical event. Different groups' interpretations are compared and discussed. Audiences can ask actors to explain their character, function, relationships, etc.

13. Photo Tracking

Similar to above. Using a pertinent historical painting or photo-graph 'recreate' it through static mime, moving mime or spoken drama. The audience or participants are asked their thoughts and may be questioned about their backgrounds, opinions and what preceded or succeeded this instant in time.

14. Thought Tracking

From a picture or a person in role (e.g. shell-shocked soldier, Neelands, 1990, p.30) children imagine the character's inner thoughts and images. These can be shown as a series of freeze frames. Alternatively, a mime is frozen before the end. Characters are asked for thoughts and explanations, or the audience can suggest them. Suggestions are discussed and criticized, building up notions of theorizing and argument.

15. Forum Theatre

An evidential history story is acted but frozen before the end. The audience decides what the actors do next and direct them by speaking their words or describing their attitudes, actions and posture. If the same event is re-run in several different ways, it builds the idea that contemporaneous events and historical inter-pretations of events have many possible outcomes.

16. Waiting Room

In the instant just before an historical event children are asked to react, e.g. write home about D Day, prepare the town hall for Queen Victoria, decide what to do as a family during the progress of the Great Pestilence of 1347–9. If anachronisms or implausibil-ities occur these can be picked up by the audience and discussed.

Chapter 5

Storied Traditions of Pedagogy

Jimmy, the linguister . . . had found his vocation as a teacher . . . his lessons were mainly story-telling and play-acting . . . he was a moralist as teachers often are, he wanted to instil a sense of community in the children so there were certain aspects that he omitted to mention or even falsified . . . People read into things their own truths and meanings. (Unsworth, 1992, pp.511, 514)

My husband is a scientist. He didn't do any history but he knows more than me because he has read lots of historical novels. (Primary teacher during staffroom discussion, June 1994)

Telling Stories: What Are Its Precedents?

A book concerning pedagogy in history needs historical perspectives, but the political atmosphere of the 1990s has been so reductive and short-term that histories of teaching have often appeared peripheral, even self-indulgent. Historical research into educational systems is relatively common but 'There are no classrooms, no children, no teaching, no learning . . . not the history of those classroom experiences that constitute an act of education . . .' (Silver, 1992, pp.103–4). Archaeology similarly has tended to ignore children in its illustrative reconstructions (Hurcombe, 1997). Such willed ignorance of past experiences — of history and theory as tools to improve current educational practice — is as foolish as it is dangerous. This brief chapter cannot compensate but does offer preliminary genealogical speculations about storied history pedagogy. We can start *before* the beginning, at least of academic history. From prehistory, a term I dislike but have included carefully, Le Goff (1992, p.58) discusses middle Palaeolithic age sculpted figures 'mythograms paralleling the mythology that develops in the verbal order.' Might such artefacts have been used educationally, alongside spoken myths and stories? As with cave paintings or flint tools, logic suggests people are likely to have learned and been taught how to construct and deploy sacred objects. As for used historically, that depends on what we mean by history:

Investigations and testimonies . . . That is the meaning of the Greek word *historia* and its Indo-European root *wid-weid* (to see). History thus began as a narrative told by someone who could say: 'I have seen, I have heard it said'. This aspect of history-as-narrative, of history-as-testimony has persisted throughout the development of historical science. (ibid, p.xvi)

Le Goff discusses other pre-literate 'pasts': collective and origin myths, genealogies and 'technical knowledge . . . transmitted by practical formulas that are deeply imbued with religious magic' (ibid, p.58). He also argues against these 'pasts' being antecedents of 'academic history' since they are not based on eye-witness accounts. Our quest is less exclusive and Le Goff's objects, myths, family trees and mnemonic formulae may have functioned as teaching aids or story props. The world-wide prevalence of story in early sacred texts and the use of props or objects in much traditional storytelling (Pellowski, 1977) all support such a speculation.

Jumping to classical Greece, Herodotus's histories would have been difficult for him to create without being preceded by personal, social historical learning, taprooted into the culture in which he grew up. This valued written records but Herodotus also used 'his own observation, inquiry, conjecture . . . rational analysis . . . and oral traditions' (Herodotus transl. de Selincourt, 1996, p.xviii). Modern Westernized and industrial constructs of education tend unnecessarily to derogate myths (Beare and Slaughter, 1993, pp.82–6). Levi-Strauss was less judgmental: 'The obstinate fidelity to a past conceived as a timeless model, rather than as a stage in the historical process, betrays no moral or intellectual deficiency whatsoever' (1966, p.236). Therefore, rather than viewing artefacts as somehow artificially separate from academic history's family tree, as does Le Goff (1992), for Levi-Strauss ancient objects may be incarnated archives. The Australian Aranda tribes' wooden or stone *churinga* (objects) are:

> Hidden in piles in natural caves . . . Periodically they are taken out to be inspected and handled . . . polished, greased and coloured . . . Their role and the treatment accorded to them thus have striking analogies with the documentary archives which we secrete in strongboxes . . . and which we inspect from time to time with the care due to sacred things . . . (Levi-Strauss, 1966, p.238)

Gravestones, memorials, statues, family stories, family trees: such recognizable social and psychological phenomena perhaps had educational functions in the past about the past, preceding modern, academic understandings of the word *history*. Now they are a fashionable element of academic studies (Times Higher Educational Supplement, 1 July 1994, pp.17–19), rarefied historical academe is debunked in academic books (Samuel, 1994), and a (female) Professor of Social Anthropology can reinvent the Greek muse of memory (Mnemosyne) as androgynous rather than female! (Tonkin, 1992, p.112). Historically, people 'tie time and place together for different purposes and in ways which show how culture-bound any one interpretation of their relationship would be' (ibid, p.127).

What of formal schooling — conscious, organized pedagogy? Homer had a seventh/eighth century BC 'didactic purpose' (Havelock, 1986, p.11). Plato in the fourth century BC thought Homer's stories needed educational and gendered censorship:

> Our first business is to supervise the production of stories . . . We shall persuade mothers and nurses to tell our chosen stories to their children, and by means of them to mould their minds and characters . . . The greater part of the stories current today we shall have to reject. (Lee, 1974, p.131, also p.437)

Because he belonged to a time that saw textbooks competing with oral education (Havelock, 1986, p.4), Plato combed the *Iliad* and *Odyssey*, excising detrimental extracts (Lee, 1974, pp.140–8). Those in direct, imitative spoken drama (mimesis) were educationally commonplace (ibid, p.149) and so suffered greatly: 'For have you not noticed how dramatic and similar representations, if indulgence in them is prolonged into adult life, establish habits?' (ibid, pp.153–4). Instead 'we shall for our own good employ story-tellers and poets who are severe rather than amusing' (ibid, p.157). Yet the oral traditions of poetry, myth and story still suffused Plato's vision: 'So let us tell the tale of the education of our imaginary guardians as if we had the leisure of the traditional story teller . . . In this education you would include stories . . . of two kinds, true stories and fiction. Our education must use both, and start with fiction' (ibid, p.131). Plato's utopia had also to contrive 'some magnificent myth to carry conviction to our whole community . . . a fairy story like those the poets tell and have persuaded people to believe about the sort of thing that often happened "once upon a time", but never does now and is not likely to' (ibid, p.181, my italics).

Coeval with Plato was Thucydides. For both men the Peloponnesian war represented a life-changing episode. In Plato's case traumas prompted utopian ideas, for Thucydides they stimulated a more-seemingly 'modern' but still narrative historical record of the events. In this Thucydides mixed mythic with historical time by describing 'events which happened in the past and which (human nature being what it is) will at some time or other . . . be repeated in the future' (transl. Warner, 1954, p.24). Thucydides's historiography 'illustrates the effect of the operation of the aporetic anecdote on the writing of history' (Fineman, 1989, p.63). But he wrote to educate 'those who want to understand clearly the events which happened in the past' and had evidential aspirations: 'I have made it a principle not to write down the first story that came my way' (Thucydides, transl. Warner, 1954, p.24). Nevertheless, speeches of which Thucydides had no experience or proof were reconstructed before being written down (Fineman, 1989, p.52). Thucydides and Plato were unsure of Homer's oral stories and so desired to invent 'a textual discourse . . . to make orality obsolete' (Havelock, 1986, p.62). Accordingly Thucydides claimed:

> Better evidence than that of the poets, who exaggerate the importance of their themes, or of the prose chroniclers, who are less interested in telling the truth than in catching the attention of their public, whose authorities cannot be checked, and whose subject matter, owing to the passage of time, is mostly lost in the unreliable streams of mythology. We may claim instead to have used only the plainest evidence. (ibid, pp.23–4)

From this discussion several ancient Greek themes assume modern significance. Disconcertingly the first is their mistrust of fictional, poetic, domestic and essentially female storytelling. Such suspicions of 'women's talk . . . haunts the whole history of the old wives' tale' (Warner, 1994, p.412). I am uncomfortable that, in this brief chapter, the male suspicions of female storytelling pervading many Greek and monastic medieval historical sources may be read as my own.

They are not. The informal educational traditions of domestic storytelling are proud (e.g. Pellowski, 1977, Chapters 1–7) even if this book's focus on more formal pedagogy and schools cannot do them justice. In such a brief sketch I have had to rely on generally well-known (and male-dominated) sources. For instance the sixth century BC female poet Sappho, despite being 'the earliest Greek educator about whom we have any information' frustratingly created no recorded 'pedagogical theories' (Smith and Smith, 1994, p.6). Even Plato's support for educational equality in The Republic between men and women has been 'misinterpreted' or treated with 'significant distortion' (Rowland Martin, 1994, p.36) by many translators and commentators. More comfortably for this book Plato was a 'curriculum developer'. Plato (and Socrates) were especially fearful of the isolated teaching of 'skills' in the curriculum sophisticatedly decontextualized from moral values by Sophist 'travelling teachers' of 'rhetoric, the art of self-expression and persuasion' (Lee, 1974, p.18). As Plato questioned the Sophist division of education from moral action, I question the self-delusion that we can teach 'historical skills' isolated from consideration of moral or narrative interpretations. Plato's pupil Aristotle (transl. Sinclair, 1962, p.300) saw 2400 years ago in The Politics 'no general agreement about what the young should learn either in relation to virtue or . . . the best life.' Similarly much modern educational muddle derives from too much discussion about curriculum detail and insufficient about virtues (see also MacIntyre, 1981, p.154). It is difficult and undesirable to divorce teaching history from explorations of morality and virtues. Thinking about teaching history and doing it are 'practical sciences' (praxis) in Aristotelian, virtuous terms:

> Informed action which, by reflection on its character and consequences, reflexively changes the knowledge-base which informs it . . . Praxis . . . constantly reviews action and the knowledge which informs it. Praxis is always guided by a moral disposition to act truly and justly, called by the Greeks phronesis. (Carr and Kemmis 1986, p.33; also Lather, 1986)

Practice in reflexively deconstructing and reconstructing stories can develop intellectual capabilities *and* virtues by asking of reality 'why did it end like this?' But as well as doubting the modernist separation of ethics and history, about which I have felt uneasy since entering teaching, I am also criticizing my own previously Whiggish ideas of history education: that present teaching was better, future education likely to be perfect and past pedagogues a stunted species. Realities are far more complex. Past societies evolved the history education that appeared appropriate to the people who controlled it. I have referred to myths as history, Homeric orality and Platonic texts. Medieval, monkish annals are another 'different kind of representation' (White, 1981, p.16) not worse history but still inescapably moral, and mostly male. In annals, events are ordered by God (ibid, p.14). In a chronicle divine sequence gives way to a human and moral translation of history:

> If every fully realised story . . . points to a moral, or endows events, whether real or imaginary, with a significance that they do not possess as mere sequence, then it seems possible to conclude that every historical narrative has as its latent or manifest purpose the desire to moralise the events of which it treats. (ibid, pp.13–14)

This argument, that as teachers we morally translate history by telling and prompting stories (Chapter 8) is confronted by serious problems. While 'stories are constructions of experiences' (Amy, 1986, p.20) narrativizing expresses a human desire for moral control over events. Such a view conflicts with many aspects of modern managerialism, whose proponents 'justify themselves and their claims to authority, power and money by invoking their own competence as scientific managers of social change' (MacIntyre, 1981, p.82). National curricula and initiatives crawl with such examples of 'scientific management of social change', displacing Aristotelian virtues concerning ends with a *morality separable from action* and controlled only by ethics (MacIntyre, 1981, p.79). Now I am not arguing that what we teach should have a didactic moral message, attempted through 'oral recitation' in some 1840s American schools (Hamilton, 1989, p.126). Here talk became a tool morally to inoculate children, the result of an intellectual mechanization of education into 'batch production' on factory principles (ibid, p.139). English equivalents abounded:

> In catechetical lessons the parroted factual responses memorized by whole classes included . . . 'dates' in history . . . and equally presented as incontrovertible truths, unfavourable stereotypes of other peoples. In Anglican schools, anti-papist sentiment was endemic . . . using, no doubt to good pedagogic effect, the power of contrast: Mary (despotic and cruel) and Victoria (liberal and benevolent). (Marsden, 1993, p.323)

Equally, such crude historical examples should not discredit the potentially rich relationships between storied oral education, morality and history. After all and contemporaneously, native North American peoples were using spoken story to teach about gender, migration, hunting skills, psychology and history (Roemer, 1983, p.42); a narrative tradition still in educational existence (McCabe, 1996, Chapter 8). This is neither Whiggishly nor WASPishly simple. Recitative teaching methods can be weak not because recitation is wrong, but because alone and in systematized settings they lack the capacity to elevate what pupils learn to Aristotelian praxis, to reflexive knowledge: their Victorian 'end' was obedience and dependence (Marsden, 1993, p.324). Yet I no longer crudely reject them simply as past pedagogy with nothing to offer. Indeed, I am increasingly uncertain that to view history as amoral is educative in the widest sense. Perhaps we can use history not to teach *a* or *the* morality, but to explore how moralities developed and to help pupils reflexively to develop their own? This is hardly a new impulse and links with nineteenth century justifications to include history in school curricula better to educate the newly enfranchised. Low-Beer (1966, pp.154–5) defined moral history pragmatically: 'not the cosmic, eternal or ultimate moral values which are the concerns of myths, religion or philosophy' but rather 'the morality of the world . . . social and public rather than individual morality. It shows us how people actually behaved.' Should we harness the excitement spawned by history's ability to raise with children moral and philosophical questions? (e.g. Matthews, 1984). 'Never', I can hear liberal friends saying, 'it would bring back those awful Victorian moralising textbooks.' Chancellor researched hundreds of these but she doubted that the:

Modern penchant for facts ... without judgements is not more truly redolent of complacency and of an unwillingness to face the implications of human actions. Victorian textbooks were full of the horror of the death of a few hundred Protestant martyrs of Mary's reign; their modern counterparts describe the suffering of millions in concentration or forced labour camps or in bombing raids, without undue emphasis. (1970, p.142)

It is an historical sin to stereotype Victorians as simplistic. The Educational Times of 1 July 1873 found the selection of history textbooks so difficult that they 'decided to adopt the oral system of teaching, preferring to leave the matter in the hands of the master to adopting any of the existing objectionable textbooks on the subject' (Chancellor, 1970, p.11, my italics). I cannot tell whether the paper meant by the 'oral system of teaching' Homer's poems, Hamilton's recitation, or something closer to my vision of a history storyteller offering interpretations as a stimulus for individualized deconstruction and reconstruction. What is evident is that morality was not distinct from history in this view, it was simply *too sensitive to leave to the printed word*, either as textbook or centrally prescribed curriculum. I suspect that inspired or radical teachers will, regardless, break central rules, discard written materials and teach through talk, as with this East End Edwardian:

Now we 'ad a man at Wick Road school ... today you'd call 'im a communist, but looking back, he didn't tell us history out of the books. Now if you got hold of all those history books we had at the time, they was all a load of flannel, about Edward the Peacemaker, Queen Victoria and Elizabeth the First. When he did give us history the way he did give it, he did show us that they wasn't as glorious as what they made out ... He gave us a truer picture because all the books were glorifying the monarchy ... (Humphries, 1981, p.43)

I ask some plain questions. Are the current and coming generations of teachers intellectually and socially confident enough to act with similar independence? Should we accept the industrialized, batch production of pupils through centrally decided curricula translated by a handful of programme makers, publishers and writers such as myself? Is history education at any level useful, if it is decontextualized from understanding of ways of making moral meaning? My present answer to all three questions is firmly *no*. Instead, I see more scope for educating teachers and children into a less centralized 'oral system of teaching' aimed not only at inducting pupils into knowledge, but also into their own constructions of historical and moral meaning. Some of our Victorian predecessors had to make similar choices, between quantitative activities resulting in accountable measurements and qualitative teaching gambling on lifelong learning.

Ilketshall St Lawrence School Log Book, June 12th 1896

Reverend Garforth called in this morning but instead of testing the registers (as they had been examined the required number of times for the current year) he gave a description of the Indian Mutiny which was thoroughly enjoyed by the class.

I cannot vouch for the Reverend's lesson: it may have been the most biased, imperialist, closed narrative ever to disgrace history education. What I can do is refer to oral traditions of educational storytelling which, had he known of them, may usefully have informed it.

Traditions of Questioning Stories?

Oral history, in my understanding, consists of experiences mediated into stories and made educational and historical by questioning. Its expressions can be formal, scholarly and historical, accounting for their own bias; but its practice often relies upon participants' informal interpersonal skills, taught through experience and socialization. Casual chats, structured interviews and performing traditions seem to me by definition informally but educatively to fuse sources into understanding: though even for Herodotus, one of the earliest oral historians 'Inquiry, it seems, is not a simple matter of asking questions and getting answers' (transl. de Selincourt, 1996, intro. Marincola p.xxviii). This matters because formal schooling has a short, patchy past. For the majority of pre-nineteenth century historical time, mass history education occurred through unselfconscious socialization, storytelling and questioning rather than through compulsory, bureaucratized schooling. Society's elders — male and female (Pellowski, 1977) would answer questions about the past (oral history) as well as tell historical stories, legends and myths (the oral tradition). Perhaps Homer but certainly Herodotus and Thucydides were oral historians (Henige, 1982; Thomas, 1990) and either intentionally or casually educative. Socrates' thinking was likewise oral and educative: 'My method is to call in support of my statements the evidence of a single witness . . . See now if you are prepared to submit yourself in your turn to examination by answering my questions' (*The Gorgias* transl. Hamilton, 1971, p.61). Similarly, he used stories and allegories (Smith and Smith, 1994, p.17) and drew ethical lessons from historical material to conclude that past statesmen 'were no better than the men of our time' (ibid, p.132). For Socrates talk, questions and memories helped education morally to contextualize the *Sophisticated* rhetorical skills taught to adolescents (Abbs, 1993; Corbett, 1990; Jenkins, 1986). This was individualized education: 'Socrates' methods . . . had little to do with class teaching' (Hamilton, 1990, pp.75–6) and bore no resemblance to the relatively recently recommended (AAM, 1950, p.61) distortion of classes chanting history questions and answers.

Romans such as Tacitus in the first century AD (transl. Wellesley, 1964, p.12) and Lucian and Herodian in the third century AD (Thompson, 1988) continued Thucydides' use of oral historical memories and interviews. Following Greek tradition, such histories were educational in a more heroic, exemplary or moral sense than a modern ear finds familiar: 'The period was not so barren of merit that it failed to teach some good lessons as well. Mothers accompanied their children in flight . . . Distinguished men driven to suicide faced the last agony with unflinching courage' (Tacitus, transl. Wellesley, 1964, pp.22–3). Oral educational and historical traditions were also strong amongst peoples conquered by Rome. Two cultural examples inspiring profuse oral traditions were the Druidic and Christian faiths. We know that some Romano–British women were communicatively and culturally literate in Roman authors such as Virgil and Ovid and perhaps in oral tales 'giving the British version of history' (Allason-Jones, 1989, p.176). Caesar described the oral pedagogy of the Druidic Gauls, who consciously rejected Greek education:

> Large numbers of young men flock to them for instruction . . . It is said that these pupils have to memorize a great number of verses — so many that some of them spend twenty years at their studies. The Druids believe that their religion forbids them to commit their teachings to writing, although for most other purposes the Gauls use the Greek alphabet. (The Conquest of Gaul transl. Handford, 1951, pp.31–2)

To an extent this was reflected in early Christian groups whose intermittent persecution until 313 AD (Cooling, 1994) and general illiteracy fostered a healthy oral tradition: 'word of mouth was necessarily the most effective and popular way of passing on Christian teachings . . . until such time as a written canon was developed' (Henige, 1982, pp.8–9). Similarly 'in earliest Islamic society orality was particularly prized . . . a cardinal principle was that eyewitness evidence was to be preferred to any form of written documentation' (ibid, p.11). This elevation of oral over written traditions inverts current scholarly attitudes, typically sceptical that 'so-called fixed texts can be transmitted over several generations' and emphasizing 'the inescapably "transactional" nature of collecting oral data . . . person to person' (ibid, p.22). This scholarly realization is not new. Bede taught 'good history' (Loyn, 1962, p.270) in eighth century classrooms by using storied, popular and local oral history in an Anglo-Saxon textbook 'copied extensively throughout the Christian world' (ibid). This textbook also qualified some of its sources:

> Because, as the laws of history require, I have laboured honestly to transmit whatever I could ascertain from common report for the instruction of posterity . . . With regard to [*Northumbrian*] events . . . I am not dependent on any one author, but on countless faithful witnesses who either know or remember the facts. (A History of the English Church and People, Sherley-Price, 1968, p.35)

Bede's teaching was moral and often anecdotal: 'I myself know a brother . . . who lived in a noble monastery but lived an ignoble life . . . much addicted to drunkenness and other pleasures.' The brother, falling ill, witnessed his allotted place in hell but died before repentance and 'was talked of far and wide' (ibid, pp.297–8). Monastic historical educators such as Bede, Aelfric and Alcuin also wrote specifically oral educational texts and Latin grammars (Hurt, 1972, p.107). Aelfric relied on orality in his lives of five English saints: Alban, Aethelthryth and especially Oswald drew on Bede and the oral tradition (Sherley-Price, 1968, pp.142–4), Edmund on Abbo's literalizing of oral tradition (Chapter 7) whereas for Swithun, Aelfric added 'Some of his own memories of the veneration . . . the appearance of Swithun in a vision . . . and how the walls of the church at Winchester were hung full of the crutches and stools of cripples who had been healed by Swithun' (Hurt, 1972, pp.78–9). Bede and Aelfric combined oral history and written tradition to produce story-led, moralizing lessons in history education closer to my own practice than I had realized (Chapter 7). Similarly Snorre Sturlason (Laing, 1930, p.3) in a twelfth century Icelandic oral history of Norway, introduced evidential scepticism. 'In this book I have had old stories written down . . . Now although we cannot just say what truth there may be in these, yet we have the certainty that old and wise men hold them to be true.' Such literature had an educational objective 'to encourage people to attain a better understanding of their neighbours and a truer knowledge of themselves, through studying the real and imagined fates of their forebears' (Palsson, 1971, p.8). Claims to education, entertainment and veracity intertwine in the prologue of the immensely popular medieval history The Travels of Marco Polo which claimed to 'set down things seen as seen, things heard as heard, so that our book may be an accurate record . . . that others . . . may learn from' (Latham, 1958, p.22). More formally Abelard, twelfth century philosopher, Christian logician and teacher of theology, made rediscovered Socratic-style question and answer a mainstay of his pedagogy (Smith and Smith, 1994, pp.71–4). Following the epic episode with Heloise, in his own words Abelard's students urged him to teach theology using 'human and logical reasons . . . something intelligible rather than mere words' (Radice, 1973, pp.77–8). This was because Abelard's former teacher Anselm 'could win the admiration of an audience, but *he was useless when put to the question*. He had a remarkable command of words but their meaning was worthless and devoid of all sense' (ibid, p.62, my italics). Like Socrates, Abelard also battled with authority through *disputatio*, a new teaching style:

> By *disputatio* is meant a new technique to replace the traditional *lectio*, a lecture by a teacher . . . read aloud sentence by sentence . . . Disputation adopted a more conversational method, posing a problem and discussing it by means of question and answer, by setting out the difficulties and attempting to resolve conflicts. (ibid, p.11)

This brings us back to modernity. For Stenhouse (1979, p.115) Abelard's thinking and teaching was praiseworthily 'public' and unlike history in nineteenth century universities. 'In place of the speculative disputation open to the student as participant observer, enquiry was expressed in the archive search or laboratory experiments, mute occupations whose meaning was not self-explanatory to the observer.' Stenhouse cited Abelard to support his vision of research-based teaching. My arguments and the substantive and methodological content of this book fit the same mould. Question and answer, disputatio in school and pedagogic history, can be encouraged by making histories storied and spoken, thus clarifying the narrative scaffolding on which hangs the structure of history. Do any modern traditions support the thesis?

Theories of History Teachers-as-Storytellers

This book cannot do justice to others' extensive theorizing about and practice of educational story, but the following summaries sketch some particularly relevant empirical or theoretical perspectives.

Applebee

Blending empirical research findings on children's stories into a framework influenced by Vygotsky and Piaget, Applebee observed that historical writing is *transactional*, conveying a type of experience with origins outside of the self (Applebee, 1978, pp.10–12). Transactional language has more explicit rules and paradigms than its counterpart *poetic* language, which is derived from internalized experiences and is subjectively tacitly and simultaneously interpreted at several different levels: 'The phonemic, semantic, syntactic and thematic structures of a poem, for example . . . can never be transactionally paraphrased' (Applebee, 1978, p.15). History stories can stimulate both types of language, for history teachers mediate for and with children between interpretations of objective experiences expressed *transactionally*, often called *fact*; and interpretations of subjective experiences expressed *poetically*, often called *fiction.*

Applebee then introduces roles. The poetic tends towards the spectator role in which we 'look on . . . but we do not rush in to interrupt — to do so would obscure the relationships and spoil the effect of the whole' (ibid, p.16). The transactional is more participative, tending to be information-rich, argumentative and dialogic. Both have rich traditions: telling history stories leads with poetic, spectator-role language and questioning stories speaks more of transactional participant-role language. Neither are mutually exclusive and many history lessons I have led or observed mixed both. Many more would have benefited from explicitly exploring with children such different roles and language types. For instance:

> We may read Defoe's 'Journal of the Plague Year' as a work of literature [spectator] or decide to treat it as a source of information about the plague [participant]. The shift this brings about is dramatic: in the spectator role we may find it an exciting tale, but in the participant role throw it away in disgust when we discover that it is not the firsthand account that it purports to be. We ask different questions of work in the two roles, bring different criteria to bear on them, and take away quite different impressions. (Applebee, 1978, pp.17–18)

Yet Defoe's work is described in a (generally excellent) primary history scheme (Vicary et al., 1993) as an 'eyewitness account'. Blurring transactional and poetic language and participant and spectator roles is disingenuous if the provenance of a text's or lesson's language is unclear. Applebee expressed the discursive end of this meaning-making process as reformulation: At the other end is articulation where meanings are confidently assigned to acts. In the middle is the pivotal educational area of the expressive. Here people share information in participant roles such as discussion, and view it in spectator roles as in biographies or travel books (Applebee, 1978, pp.22–3). Oral expressive language is particularly educationally rich: 'talking through . . . is at the heart of the expressive, and out of which the more differentiated forms of language evolve . . .' (ibid, pp.26–7). I wish I had read Applebee during my PGCE course: it has taken me years to empirically discover that this applied to my own teaching.

Similarly valuable is Applebee's exploration of story's binary opposites or extremes (also Egan, 1989, pp.26–31), linked to children's changing perceptions of fact and fiction:

> The earliest interpretation seems to be that a story is something that happened in the past, a history rather than a fictional construct. This . . . belief in the immutability of stories . . . is eventually shaken by the recognition that behind each story there is a human author who has made it up. (Applebee, 1978, p.38)

In the next stage and borrowing from Piaget, children start to view the world of story as malleable because it is fictional. 'It is only after the story has emerged as a fiction that it can begin a new journey toward a role in the exploration of the world not as it is but as it might be, a world which poses alternatives rather than declares certainties' (ibid, p.41, my emphasis). Applebee's empirical observations of children learning link to my own treating of educational history as information-rich fiction with which to help children explore uncertain actual and historical worlds (Chapters 2 and 3). It is not that history is inconsequential, as a mere plaything or idle story. Rather that its fictional element opens it to the making and recreation of knowledge in young minds. I imagine an historically fictional explanation: now are there any facts to destroy or support it?

Claire

What distinguishes Claire's (e.g. 1996) work is the creative and critical ways in which she uses stories to promote the ethics of equality and diversity. Her practical examples contain valuable sketches of how this can be achieved, taking individual aspects of NC history and resourcing them with storied and inclusive teaching ideas. The local-global nature of Claire's and Collicott's (e.g. 1986; 1992; 1993) storied history has helped to sustain my own curriculum development ideas, and may resonate with readers. Is one reason for the increasing popularity of local, community and family storied histories the globalization of our culture through world trading, travelling and communication systems? That as the pace of change speeds in some areas of our lives, so we feel an urgent and compensating need to be rooted in others? Such phenomena may also explain why I am writing and you are reading this book.

Egan

Egan argues that teachers transmit cultures and values: 'The teacher's job is to help children in a few years to gather to themselves the achievements whose first accomplishment took centuries and millennia' (1979, p.106). Teachers 'can be seen as the story-tellers of our tribe . . . the curriculum is the story they are to tell. The art of teaching is . . . tied to the ancient and powerful tradition of storytelling' (Egan, 1989, p.109). To support this he developed theories of educational or child development (1979, 1983) as well as of curriculum development (1985, 1989, 1990). The former outlines four approximate 'educational' stages of development through which children proceed but never 'leave': mythic (up to about 8 years old) romantic (perhaps 8–15) philosophic (maybe 15–20) and ironic (20 to adulthood). These mirror educational and literary, rather than psychological or social scientific research. Egan celebrates emotion, imagination and morality as elements in or aims for teaching, and argues that the story form has a central place in achieving them. His is an alternative view of education in contrast to the behavioural, the social scientific and the quantitative:

> It is a model for planning teaching that encourages us to see lessons or units as good stories to be told rather than sets of objectives to be attained. It is an organic approach that puts meaning centre stage . . . it is about how to use the power of the story form in order to teach any content more engagingly and meaningfully. (Egan, 1989, p.2)

Like others I find this model inviting, sustaining and inspiring (e.g. O'Donoghue and Saville, 1996). It also raises important questions (e.g. Buckley, 1994) to which I add:

1. To put 'meaning centre stage' begs the questions *whose* meaning, and who decides it? Teaching children and communities to make their *own* meanings, decentralizing the power of myth, story and history-making brings with it a concomitant challenge to existing structures of power.
2. Egan (1989, p.108) sketches a 'Great True Stories of the World' curriculum. The word *true* suggests orthodoxies, not interpretations. *Real* might be safer since it connotes actual things, but leaves pupils to decide *truth*.
3. Are the meanings of well-known stories or histories 'obvious' (Egan, 1989, pp.56–61)? Do stories necessarily mean the same thing to teachers, teenagers or infants?
4. Egan (e.g. 1989, pp.33–8) is suspicious of objectives-led curricula, as am I. Years of classroom experience also inform me that most children like clear objectives and constructive tasks. Might objectives help children make meaning of education — and so be part though not all of the story? (See also Stenhouse, 1975, p.80).
5. For Egan 'we learn some of the most important, profound and powerful concepts very early, and thereafter we refine and elaborate them' (1989, pp.44–5). I am unsure. Might children become fascinated by emotional concepts such as fidelity (Egan, 1989, pp.54–5) because they do not yet know them, but observe their importance in the adult world?

Fines

We explored Fines' views of teaching, history and story in Chapter 2 and draw on his advice for telling stories at the end of Chapter 7. Here two other recent texts are explored. Both draw from a lifetime's teaching and recent extensive experience of teaching NC history across England during the Nuffield History Project (Fines and Nichol, 1997, p.viii). Few writers about education have so extensively or successfully structured their writing and theorizing around recent, practical pedagogy. This book shadows such an approach even if it cannot emulate it.

Fines' purposes of story in history classrooms either excluded some of those I summarized at the end of Chapter 2, or simplified them out. But — slightly paraphrased — they centre on how stories can assist *learners* by:

- conveying information;
- creating contextualized products;
- fostering imaginative wonder;
- relating the past to the present. (Fines and Nichol, 1997, pp.185–6)

In preparing stories from the *teacher's* perspective, Fines excludes plot and emphasizes simplicity (ibid, p.188, see also this chapter's **In Practice** section). He also argues (Fines, 1997, p.53) that, in the hands of experienced practitioners, story in history can be approached thus:

1. *Encouraging children into enchantment* through tasks linked to a story: observing, spacing, arranging, talking etc.
2. *Teachers explicating the subtexts* of stimulating tasks so that children learn consciously as well as by osmosis. At this stage the story's objectives predominate because convincing imaginative frameworks encourage children to read evidence better than do bureaucratized learning targets.
3. *Thickening through description*, characters, places, objects and events about which we can ask questions.
4. *Leading children to learning* through outcomes both as resolutions of the problem of a story, and as curricular or personal educational aims and objectives.

Such a model of planning derives not only from deeply grounded experience, but from values that place children's imaginations, alongside cultural riches, at the centre of a curriculum: two hoards of treasures made accessible to teachers through story.

Gudmundsdottir

Gudmundsdottir investigated American 'teachers as storytellers' by modifying Applebee's ideas and using them, during extended observations, to compare experienced and inexperienced practitioners. Thus a novice *Chris* had 'good stories for the units where he is able to draw on his content knowledge . . . however, he is unable to maintain his stories across units'. In contrast experienced teachers' curriculum stories are like narratives, with central ideas shaping and moving plots. These experienced teachers' stories helped them 'Move between individual historical events, major historical ideas and classroom activities and assignments. The stories represent a unique interaction between their personal theory of the teaching of history and the practice of teaching' (Gudmundsdottir, 1990, pp.116, 117).

Gudmundsdottir claimed that 'Curriculum stories help teachers manage complex ideas and make them accessible for students . . .' (ibid, 1990, p.117). Citing two history teachers, she argued that storytelling did not so much simplify problems as extend them to include considerations of moral quandaries (Gudmundsdottir, 1991, p.216). Examples cited English, science and history teachers. Buttressing this book she sketches narrative's power to encapsulate experience, interpret classroom materials, encourage reflection and transform teachers and learners (Gudmundsdottir, 1995). Such 'anecdotal pedagogy' is powerfully represented in my own life history by fond memories of inspiring and eccentric history teachers. But how many boring, partial or self-indulgently anecdotal history teachers could we also recall? Critical analysis of teacher talk reveals its dangers as well as strengths (e.g. Barnes, 1976; Hull, 1986). Nevertheless the appeal to 'consider teaching as a timeless text, a continuing tradition of stories told and retold with the express purpose of engaging students' (Gudmundsdottir, 1995, p.35) seems as relevant to our age in general, as to this book in particular. Such an approach also has its dangers, as the next chapter highlights.

In Practice (5)

Preparing Stories for Classrooms

How can you start to story your curriculum? Fines and Nichol (1997, p.188) exclude plot and emphasize extreme simplicity:

- knowing what the story is about and for;
- identifying the shape of the story — its problem and solution;
- thickening descriptions from a child audience's point of view.

This is invaluable advice for preparing everyday teaching, but told or written stories also function across a wider range of timescales. Stories can lead lessons (short term), hold together units (medium term) and link across years (long term). Without losing five minute stories, consider longer ones too as in the following example.

A Life in Tudor Times: A Storied Approach to Planning

Preparing stories	Short-term lessons e.g.:	Medium-term units e.g.:	Long-term across units e.g.:
1. Select enticing evidence of a human question for the story's heart	Would you want Henry VIII as a father or friend?	What stories lie behind Tudor: paintings, poems, buildings, letters, music, diaries and books?	How did Tudor people judge a hero or villain? How do their judgments compare to ours or those from other times?
2. Identify central ideas, themes and evidence	Henry's story told through the eyes of a servant (1510s), Anne Boleyn (1520s), Thomas Cromwell (1530s).	Which words matter in the Tudor stories we study? (e.g. kingship for Henry). Which words might Tudors have used?	About which Tudor life stories or events can we find out and how? What are their themes? (e.g. faith, peace, wealth, invasion).
3. Research key points to structure the story and key descriptions to clothe the narrative	Henry's handsome image at the accession, courting and discarding Anne Boleyn, the work and death of Thomas Cromwell.	Henry VII, Henry VIII and his family and ministers, Francis Drake, Philip II and the Armada, William Shakespeare. Local characters, e.g. criminals, traders, farmers, martyrs.	Often identified by NC definitions, e.g. the break with Rome, exploration, the Armada, court, town and country life, arts and architecture.

Preparing stories	Short-term lessons e.g.:	Medium-term units e.g.:	Long-term across units e.g.:
4. Weave in historical certainties and uncertainties	Was Henry ever truly handsome? What if Elizabeth had been born a boy? Of what was Cromwell guilty?	Focus upon key uncertainties in some stories: e.g. Francis Drake — brutal pirate or heroic genius?	Can we ever know all of a story? How have stories changed over time?
5. Devise how children can reflect upon the story	Children discuss and decide these points at the end of each mini-story.	Children choose a Tudor life story to research, communicate and judge as 'experts'.	How can we be fair to people who are dead and gone? Children experiment with making judgments.

Planning for Stories

1. Select Enticing Evidence of a Human Question for the Story's Heart

To gather momentum history needs evidence, stories need tension and children need motivation. Problems are their fuel, embodied through questions: not just history teachers', but questions asked at the time by characters or arising from children's curiosity. Such questions can start a story abruptly and/or accrue. Endings resolve the tensions created, though as this is history and not literature, avoid the simplistic 'happy or sad'? We may finish by answering a story's central problem or question, but leaving another for children to consider.

2. Identify Central Ideas, Themes and Evidence

Any story contains many of these, but a teacher's job is to choose those that can be coherent and significant from a learner's point of view. Perhaps one major theme and a couple of smaller ones avoid confusing younger learners. This does not preclude returning to a story to highlight different ideas or evidence. Central ideas can be:

- decided by you in advance;
- judged by children's reactions to preparatory work;
- highlighted during a story based upon children's reactions;
- decided by children after telling: what does the story mean to *them*?

For instance, the story of Henry VIII offers many themes relevant to children: glamour, hope, advice, faith, bullying, etc. Story also domesticates thematic historical concepts (e.g. invasion, economy, trade), making these meaningful by showing children how such ideas describe past events at life pace and with unforeseen outcomes.

3. Research Key Points to Structure the Story and Key Descriptions to Clothe the Narrative

Storytelling involves knowing a plot, often by dividing the story into manageable periods of time, e.g. Henry's marriage to Anne Boleyn or important events, e.g. the sinking of the Mary Rose. This involves teacher-judgment even within a central curriculum, made historically safer by emphasizing the ideas of evidence and uncertainty.

4. Weave in Historical Certainties and Uncertainties

How reliable are historical 'facts'? For instance, during or after telling divide a story into things about which we are certain and uncertain, perhaps with an additional column for questions to ask. Always refer to the central role of the teller — this is one person's interpretation, not simple 'truth'. Once certainties have been logically and chronologically arranged then uncertainties can be examined. Then discuss how some apparent certainties (e.g. Elizabeth I seems fine-looking) may be simplifications or stereotypes open to reinterpretation.

5. Devise How Children Can Reflect upon the Story

Teaching strategies to support children's reflections are detailed at the end of all following chapters and discussed in the main body of the text.

Chapter 6

Why Stories Need Critics

I know from assembly how children react to stories. You tell them some facts and only half are listening . . . You tell them a story and they're all there with you, wanting to know . . . it's the same whether it's real or made up. (Infant school headteacher, September 1994, discussing curriculum plans)

Learning does not consist only of knowing what we must or we can do, but also of knowing what we could do and perhaps should not do. (Eco, 1980, p.97, William talking)

Some Outcomes and Principles for a History Curriculum Based upon Stories

This book speculates about rediscovering traditional ways of teaching and learning history by presenting the curriculum as stories waiting to be heard, read, interpreted and reconstructed using evidence. Such a compromise between the new and traditional tries to root itself in teaching practice and in the history of history education. This avoids the trap of exhorting teachers to behave in radically different ways (Goodson, 1978) while valuing educational gains from 'new' and NC histories. The resultant curriculum is almost recapitulative, a version of history in education based upon where history began: the telling of stories and the questioning of oral history. The need for it has long been recognized:

Perhaps something has been lost of that wondrous fascination which a good story from the past exerts over children. I must confess that I am growing a little tired of using national curriculum texts which line up endless pages of sources like hurdles which have to be clambered over before getting down to a bit of the story of history . . . (De Marco, 1992)

Based upon practice, testing and research I therefore offer some 'teaching principles' and 'learning outcomes' for educational history using stories. Principles of educational procedure were postulated by Stenhouse (1975, pp.87–94) who also recognized the necessity to synthesize:

Instruction, which gives us access to conclusions which represent in simplified and hence distorted form our best grasp of a realm of knowledge and meaning; and *learning by enquiry or discovery*, which enables us to understand how to utilise such a representation of knowledge, to assess its limitations, and to develop the means of pushing outwards beyond these limitations. (Stenhouse, 1982, p.118)

I adopt Labbett's (1995) model of 'should . . . rather than' for principles, reflecting how teachers incessantly choose between attractive courses of action. Principles matter for: 'When we tackle story as being historical . . . we are presumably putting limits on what is seen as acceptable. Whereas if we are just talking about stories in the tall story direction then anything goes' (John Fairclough, Museum Education Officer Interview, April 1993). The following are offered for use within any educational context mixing story and history but are *advice to test not to follow.*

1   A Story's Purposes and Values

Teaching principles

1. allow for sheer enjoyment of a story rather than always attaching aims and objectives;
2. enact and call whole-school values and policies to account rather than take them for granted;
3. teach literacy embedded within rather than decontextualized from genuine and interesting texts and purposes;
4. extend rather than curtail learner autonomy;
5. help students consider meanings and moral implications rather than assume they are obvious;
6. stimulate talk or thought rather than leave listeners inert;
7. embed within rather than disembody history from wider curricula;
8. individualize rather than standardize outcomes by building on learners' prior knowledge and current response.

Learning outcomes

1. learners enjoy a story for its own sake;
2. learners think about valuing cultural diversity or equal opportunities;
3. learners practise and improve aspects of literacy;
4. learners are more confident or independent for having listened to a story;
5. learners discuss the meanings of stories;
6. learners are sufficiently motivated through a story to respond to it;
7. children make links with previous learning or other subjects;
8. a range of responses encompass the spread of previous attainment and current attitudes in a group.

2   A Story's Evidence and Enquiry

Teaching principles

1. build arguments from historical evidence rather than fantasy;

Learning outcomes

1. children learn how to read different types of evidence;

2. open stories' provenance and authorship to inspection rather than acceptance;
3. expose prejudices about a story rather than assume neutrality;
4. open rather than obscure the telling's modernity;
5. leave pupils inquisitive rather than accepting;
6. advance rather than assume children's understanding of historiography;
7. question historical representations rather than present stories as simply 'true'.

2. learners explain something of where a story comes from;
3. learners discuss and record different ideas about a story;
4. learners understand how a story is influenced by the time in which it is told;
5. learners ask new questions and offer new explanations as a result of a story;
6. learners increase their knowledge of history thinking and methods through a story;
7. learners understand or vary a story through other forms, genres or endings.

3   A Story's Substance and Content

Teaching principles

1. communicate memorable content and structures rather than only developing 'skills';
2. forward history as a way of thinking rather than a mass of information;
3. explain rather than assume understanding of the terms within a history story;
4. acknowledge rather than ignore a story's affective elements;
5. refer to rather than simplify out gaps, imbalances or uncertainties;
6. communicate contemporaries' range of choices and disagreements rather than project historical outcomes as consensual or inevitable;
7. discourage rather than promote stereotypical images and understandings.

Learning outcomes

1. increase the knowledge of history and stories in learners' long-term memories;
2. children become more skilful in their use of major historical concepts (e.g. chronology, change, continuity, comparison);
3. learners understand more of history and story's vocabulary;
4. learn from and with emotions raised by a story;
5. learners see a story's flaws as well as strengths;
6. learners appreciate how characters often choose between courses of action;
7. children learn through history that humans can behave unpredictably.

4   *A Story's Telling*

Teaching principles	*Learning outcomes*
1. sometimes slow the pace for listener analysis rather than proceed uniformly;	1. learners offer questions or feedback within a story;
2. encourage pupils to reveal understandings to scrutiny rather than leaving them unexamined;	2. learners pose their own questions and tell their own stories about a period, person or event;
3. proceed towards aims rather than be side-tracked by idiosyncratic questions;	3. learners understand that individual questions and interruptions are of varying relevance to a wider group;
4. mirror the audience's needs rather than teller's preferences;	4. all learners understand something of what the story is trying to tell;
5. ask participants critically to evaluate rather than accept historical explanations and descriptions;	5. learners suggest weaknesses in or improvements to a telling;
6. monitor the values and meanings in words rather than assuming that language is value-free;	6. children learn how to identify words or phrases expressing prior bias or belief;
7. protect diversity of opinion rather than promote consensus (Humanities Curriculum Project, 1970 p.1).	7. learners value open disagreement from a story as well as closure and conclusion.

Some Dangers of Teaching History Through Story

Important warnings should accompany narrative texts and teacher anecdotes:

> The sum total of my history recollections at Primary school is two . . . tatty 'Unstead' style text books from which we copied huge chunks of incomprehensible and tedious text and . . . my teacher who had grown up during the Second World War, and on many long afternoons would recount numerous tales of hardship and deprivation. Needless to say, I found history BORING! . . . Children now . . . express their opinions, find out for themselves and experience primary sources. Children like to be involved and active, not mere receivers of dictated notes. Long live history! (Ann Tingey, Brooklands Primary School, September 1994)

Because teaching involves managing intellectual risks rather than avoiding them, we shall name and explore some dangers to accompany the teaching principles of Section 1.

Danger 1: Singularity

Although historians properly question basing an account on just one eyewitness, preferring diverse sources, ironically it is often their fate to synthesize these into a singular version. Singularity is a special danger in school history, with children and teachers sometimes becoming fixated by a particularly forceful historical story. Occasionally they are also ignorant or slothful. When this happens, rather than being *part* of history the story appears to children as *history itself.* Strategies to avoid such singularity include:

* halting action for analysis;
* introducing conflicting stories from other cultures or viewpoints;
* encouraging questions;
* clarifying the story's provenance;
* symbolizing conflicts of evidence and interest by highlighting different characters' views of events;
* encouraging disputatious analysis and talk.

Danger 2: Over-simplification, Glossing and Smoothing (see also Chapter 8)

Narrative histories run the danger that audiences are presented with a *fait accompli*, access to a version of the past only on grounds of acceptance. This is what I mean by over-simplification (the French revolution was caused by a shortage of bread), glossing (the Romans retreated from Britain) and smoothing (the Industrial Revolution resulted in higher standards of living). A well-told history story may be as simple, glossy and smooth as an issue of *Country Life*, but it is only educationally well told if it supplies the audience with the means critically to read it.

'Good' teachers know when to simplify and when to complicate knowledge (e.g. John, 1994a; Wragg, 1993, pp.135–6). Yet children experience and learn through simplified historical representations before and beyond schooling through popular, commercial and children's culture, including children's cultures of toys, games, television and story (e.g. Hilton, 1996; Samuel, 1994), not to mention their own ideas (Morris, 1992). School history exists not to start historical education but to continue it on a more scholarly footing. History can likewise be seen as a simplifying, narrativizing process even for professionals. They cannot include 'all the facts in the story' and choose 'not merely on logical grounds but on the basis of appropriate rhetorical strategies' (Hexter, 1972, p.251). Simplification occurs because history and education are necessarily selective attempts

> To represent historical reality in a narrative. Words are different stuff from reality
> . . . The attempt to better reflect reality in words is . . . the struggle of all scholars
> . . . All our laws of physics and generalisations in history are attempts thus to
> simplify the world without falsifying it. (Egan, 1979, pp.104, 159)

Children rely upon simplification, but unfortunately 'teachers and textbook writers are often mistaken in the ways in which they set about simplifying knowledge' (Shawyer et al., 1988, p.215). Analogies need to be familiar and 'talk' targeted: but after analysing lessons, Hull (1986) discovered that history teachers' talk often slenderized knowledge into mechanistic study skills and ignored students' contributions. Analogous trends in school history textbooks have been noted (McAleavy, 1994; Wishart, 1986).

Teachers need to help learners see historical events as 'lived at life rate' but 'occurring in situations where the outcomes still seem uncertain and decisions have to be made' (Edwards, 1978, p.63). Such suspension of disbelief necessarily entails history teaching sharing some of the imaginative, simplificatory traditions of drama or literature. But by what critical criteria can we judge historical over-simplification, glossing and smoothing? It is easier to see faults in others than to succeed oneself:

> This opening incident needs to dramatise the contribution the monks were making to civilisation and the wanton destructiveness of the Vikings . . . The Vikings raided without warning. They were berserk fighters; fearless themselves and merciless to others. Women and children were slaughtered as the Vikings burned and destroyed everything they could not steal. (Egan, 1989, p.57)

Accuracy seems imperilled here, especially through the adjective 'wanton'. What of *balance*? The Vikings have contributed sagas, myths and distinctive art and design to European civilization. *Respect for sources?* Might the account damning the Vikings have been monastic in source? Worse examples of my own similarly flawed attempts follow in the next chapter.

Danger 3: Partiality and Idiosyncrasy

This book views teaching as a morally responsible task therefore involving decision-making that is necessarily instantaneous, sensitive and local. Does central political control of education mean that England and Wales have an official history curriculum? Some thought so, seeing 1991 NC history as Tory and *partial*: 'traditional, British "national" values . . . protected by a history syllabus centred on the "facts" of the British past' (McKiernan, 1993, p.50). Numerous critics bear depressing witness to the continuing dangers of prejudice and imbalance within historical curricula and texts (e.g. Apple, 1993; Pankhania, 1994; Tomlinson, 1990; Weiner, 1993). Yet before national curricula Barnes described idiosyncratic history lessons based not upon 'publicly established systems of knowledge but upon quite trivial preconceptions set up arbitrarily' by teachers. The result was atomized, random curriculum design 'especially in subjects such as history or geography' with 'an informal ad hoc collection of knowledge which will have no particular validity beyond that lesson' (1976, p.179). A history curriculum centred upon storytelling and making might be similarly idiosyncratic or as subtly partial as official NC history. Claire (1996) offers inspiring, grounded historical stories avoiding these dangers, but to counteract personal bias it may also be necessary to approach storying systematically within a flexible national curricular framework. Meanwhile, ponder the impossibility of objective history: 'The study of memory teaches us that . . . As the source is created . . . the subjectivity is already there . . . the subjectivity of the historian is bounded not only by the data he has, but also by the subjective interpretations of the witnesses' (Vansina, 1980, p.276).

Danger 4: Propagandizing and Moralizing

Though these dangers are well-known in *all*, not just *storied* history, the role of the narrator and the necessity to end stories give special openings for moralist or propagandist bias (Baldwin, 1994, p.29). Nineteenth and twentieth century reminiscences and textbooks offer numerous imperial examples (e.g. Tomlinson, 1990): stories frozen by print and prejudice into fixed moral tales. Little Arthur's England typically shows storytelling bending past and present to a preconceived, paternalistic moral plan:

> Now, when God allowed the Romans to come and take part of the country of the Britons, and to make servants of the people, He put it into the hearts of the Romans to teach the Britons . . . the Romans built some schools, and had school masters to teach their children to read and write . . . how glad their fathers and mothers were to see them so improved. (Callcott, 1878, pp.9–10)

Children's Stories from English History (Nesbit, pp.7–8) combined Callcott's moral flavour with nationalism and imperialism, as Captain Edric Vredenburg introduced simple Britishness — or was it Englishness?

> Today the King of England rules over the greatest Empire the world has ever seen . . . Now the stories in this book . . . are true tales of some of those Royal Children . . . remember and consider the glories of this country, and how every child can help to keep it great and glorious by leading a true and straightforward life, by being obedient to those in authority over them, and by honouring their present King.

It is anachronistically easy to sneer, but I am often as tempted by moralizing from history stories as was this fine teacher: 'Human nature, as it stands, makes mankind naturally . . . selfish . . . Both the Romans and Hitler were obsessed by power and were too ambitious . . . The moral is to accept your strengths and limitations and plan your life within those' (Blyth, 1988, p.2). Such heady mixes of story, morality and history pervade our media. In the dog days of the 1990 Iraqi war, Ignatieff argued that the unthinking use of the Hitler analogy had mistakenly allowed British people 'to think of Kuwait as Czechoslovakia'. He described a mood so close to my own that I kept this cutting:

> The real reason for the sullen 'let's get on with it' mood is that we're bored. Television news, which cannot help rewriting history as melodrama, has shortened all our patience with the slow pace of real time. Television flies us to history's front row, and then, when history refuses to behave like a soap opera, we inevitably begin to cry: Where's the action? (Ignatieff, 1990)

Mistakenly analogous history stories are dangerous if ubiquitous, but a hazard darker than clumsy moralizing lurks. Namely, the delusion that *amoral* stories are possible, and that authors can divorce conclusions from values. When history stories lie unexamined, learners are denied the opportunity critically and democratically to decide for themselves *what this history shows*. Their experiences, constructs and beliefs are marginalized in favour of the often hidden assumptions contained within textbooks', televisions' or teachers' values. For history offers numerous chances to express:

> Political, moral and religious ideas, and since these are embedded in a traditional
> and often emotive story, they are arguably more open to acceptance and less liable
> to detection by the pupil. These opinions readily form part of the patterns of
> thought which develop in the individual child as well as in society as a whole.
> (Chancellor, 1970, p.8)

Chancellor was writing about Victorian textbooks but could have been talking of this modern history teacher and peace activist's pedagogic interpretation of the American Civil War:

> David's personal values . . . influence his selection of events . . . He covers only
> one battle . . . the idea he wants to highlight is that war is a bloody affair . . . The
> end of David's curriculum story is about the nuclear arms race, Star Wars, and
> superpower potential to blow up the world. (Gudmundsdottir, 1991, p.213)

NC histories, with their jargonistic and technical paraphernalia, have tried to distance themselves from such moralizing history stories. This distancing may have been as dangerous as the stories themselves. It tried to turn teachers into technicians delivering knowledge, not reflexive and responsible artists translating meaning. I am sceptical that such a view of pedagogy can improve learning, however hard it pushes to standardize it. More importantly, the notion that a body of 'amoral historical knowledge' exists which can be 'taught' in a value-free manner might leave children more, not less open to manipulation. If children have never felt the power and tug of history stories in a safe environment created by a knowing teacher, if they are not taught how to deconstruct and reconstruct real stories, how can they resist the storied or mythic televisual blandishments of political and commercial dictatorships? Just as sixteenth century Native Americans succumbed to previously unknown and therefore irresistible diseases (Wachtel, 1971, pp.93–6) so citizens' immunity to propagandizing or moralizing history stories may derive from critical exposure to, rather than exclusion from, a range of historical narratives.

Danger 5: Heroics

Carlyle's writings have come to epitomize, in modern eyes, much of what is dangerous about the heroic in history. For instance he contrasted 'brave old Samuel' [Johnson] with Rousseau 'not what I call a strong man . . . He had not the talent of silence . . . which few Frenchmen, or indeed men of any sort in these times, excel in' (Carlyle 1841, 1993 ed. Goldberg). Modern heroic contrasts are generally more subtle. Winston (1996, p.109) used contemporary children's book idealizations of Native American culture to warn of the dubious ethics of 'appropriating stories from colonized cultures' and presenting the historical 'dispossession and demise of the Indian as inevitable' (ibid, p.115). Pertinently for our purposes, Winston links a modern author's well-provenanced historical tale to 'Rousseau's belief in a primal age of innocence . . . uncorrupted by the tyranny of modern civilization' (1996, p.114). The rather idealized relationships between Native Americans and European settlers in children's historical fiction (e.g. Dupasquier, 1988; Gerrard, 1996) endorse his argument. Carlyle and Rousseau have come full circle and as I argue below, some fictionalizing of Native American history shows *the heroic* dangerously at play.

Danger 6: Presenting Fiction as Fact

Brother Eagle, a children's book relating the supposed environmental views of the nineteenth century Native American Chief Seattle was an American bestseller in 1991–92. Unfortunately Seattle's 'speech' as quoted was apparently bogus, the result of a fictionalized retelling by a screen writer in 1972 who admitted 'I did not check the historical accuracy of anything I wrote' (*Readers Digest*, August 1993, pp.53–7, also Winston, 1996, p.120). From the opposite wing it has become commonplace to deride progressive historical pedagogy as *fictional*. In the *Daily Mail* of 20 October 1992 McGovern, newly appointed SEAC history adviser condemned a BBC schools history programme for using excerpts from Monty Python's *Life of Brian*: 'It is yet another example of how fiction is being taught in schools instead of fact.' The *Sunday Times* (13 June 1993, p.11 'Sabotage of a Reform') believed in 'the education establishment's hijack of the curriculum' and that 'The history group decided that empathy — being able to put yourself in the place of historical figures — was more important than the history of Britain.' In similar vein and with no supporting evidence Phillips (*The Mail*, 17 September 1994, p.8) criticized teachers who 'knew that teaching children historical facts was wrong because it denied the higher truths of subjective creativity'. In tandem, conservative pressure groups have beaten similar drums. Deuchar (Campaign for Real Education vice-chairman) sneeringly condemned a teacher for telling KS2 children a local Tudor legend: 'the point of the exercise was to bring home to the children that the truth or falsehood of the story is immaterial; what matters is that "people believed it"' (*TES*, 12 March 1993, p.5). Deuchar fails to portray such work in a bad light. Any contemporary priest, politician or salesperson would confirm, and history furnishes numerous examples, that what people believe *is* often more important than, and different from proven factual reality (e.g. faith in an afterlife, advertising, or Stalinist collectivizations). This is not an excuse to teach children that things are true when we know or suspect otherwise, and oral history shows how totalitarian or sloganizing stories are inimical to education (e.g. Chang, 1991); but in my experience most teachers, unlike the propagandists or journalists just quoted, do not over-simplify the complex relationships between fact, fiction, imagination and information. This tape-recorded discussion about children's increasing but fragile understanding of the area is typical of many primary schools I have worked in:

KS1 teacher: I read a book which was a semi-fictionalized account . . . and the children found that so hard. One or two of the older ones could grasp . . . I almost sort of steer clear of it because they can understand the difference between *this is true and this is a made up story*, but that sort of area in-between . . .

Headteacher: By the age of nine they can begin to understand that . . . I talked quite a lot with yours about myths and legends and at nine they were beginning to understand the difference between those and truth. (March 1993, rural 5–9 primary)

Cooper (1992, p.136) usefully warned against children writing fictional history stories in which 'imagination is not necessarily tied close to evidence, and interpretations of evidence do not have to be argued as they do in discussion.' I warm to the practice of introducing more critical talk into storymaking. The existence of *fictional* or *factional* material in history education is not the real danger; rather that it is not used often, or sensitively, or skilfully enough to alert children to the dangers of history's manipulation by storytellers more unscrupulous than themselves. Pondering the Balkans of the 1990s, Smith described historical fictions metamorphosing into powerful myths as:

The defining criteria of the nation itself . . . the nationalist myth represents a break with the past, even as it recasts it. Firmly tied to a vision of the future, it seems to impart a sense of destiny and mission. Its use of the past is therefore selective, singling out foundation myths and golden ages, and omitting unworthy episodes and embarrassing interpretations. (*Times Higher Educational Supplement*, 8 January 1993)

Danger 7: The Absence of Problems or Questions

'History writing is not storytelling but problem-solving' (Fischer, 1970, p.xii). Need they be contraposed? Fine stories ask and resolve conflicting analytic, communicative, moral and artistic questions. Fischer's criticisms may apply more accurately to *descriptive* rather than *storied* history, though these can overlap. Similarly a 1991 review of Ginn History KS1 storied materials was 'doubtful whether stories of complex events, told in necessarily simple language to young children who are unable to understand adult motivations, are the best way to introduce them to the processes of historical thinking and understanding' (Cooper, 1991). Perhaps not, but might the fault lie in a lack of appropriate questions, content or evidence rather than in storied structures? Naidoo (1992, p.54) cited how a KS3 pupil thought a book about Nazi Germany, 'doesn't look very interesting'. After three chapters 'nothing makes me want to read on'. But after seeing a TV programme and questioning two elderly Jewish survivors 'their answers moved me very deeply . . . I've changed my mind about the book.' Stories and questions also need to be *plural*:

> Just one account would be hopelessly unhistorical, artificial and sterile . . . Once it is accepted that knowledge implies understanding of the grounds for historical claims, the problem ceases to be one of constructing an anodyne all-purpose account, and becomes one of enabling children to understand why it is possible to have different accounts, and how those accounts must meet certain standards. (Lee, 1991, p.49)

Danger 8: Lacking Provenance

Sophisticated media complicate the provenance of texts. For instance 'feedback' (Henige, 1982, pp.81–5) influences modern oral histories, by overlaying people's memories with media stories: interviewees literally cannot remember where their memories came from. Many children's history stories similarly obscure provenance. In Moses' (1994) excellent educational collection of *Stories from the Past* the provenance of most are unclear, there is little primary evidence and introductory notes are cursory. The judgment seems to have been made that stories need little pedagogic application or explanation. A similarly useful collection (MacDonald and Starkey, 1996) contained more background information, enabling the teacher to answer or help children construct sharper questions: but again few stories encapsulated ideas of provenance or multiple interpretations or evidence. Unaccountable transmission rather than active analysis can dominate such texts, though this may reflect a broader and continuing difficulty teachers have with primary sources. Worryingly, one empirical study of KS3 history classrooms showed secondary sources such as information and worksheets (42 per cent), diagrams/maps (18 per cent) and textbooks (10 per cent) being used nearly five times as often as primary sources (15 per cent) (John, 1994b). Since historical fiction conforms to different, though related criteria (see following) it may be unfair to criticize it on the same grounds: but surely historical stories and illustrations would be enhanced, rather than undermined, by telling readers their provenance? For instance, I have taught successfully through two titles describing nineteenth century American emigration (Dupasquier, 1988; Gerrard, 1996) but needed much more information on the stories' origins and content. Although this can partly be overcome by encouraging discussion, close reading, lingering over illustrations and comparing with non-fiction texts, nothing can replace an authentic authorial explanation.

Danger 9: Literary Pressures Undermining Historical Accuracy

Following extensive critical reading, Little and John, 1986 (also King, 1988) attributed powerful and educational historical fiction for children with characteristics equally important to teachers-as-storytellers:

- accuracy derived from research;
- subtle scholarship unmarring momentum;
- linguistic and psychological accessibility;
- authenticity to the characters and times in question;
- personally and historically credible dialogue;
- balance between universal dilemmas, literary themes and our knowledge of the period;
- a sense of historical uninevitability conveyed by characters.

Such discriminations are vital in classrooms, where stories can smother the ideas they were meant to spawn. As a colleague commented of her 1970s practice:

> I mainly taught 4- to 6-year olds and the history we did . . . evolved from anniversaries, e.g. Guy Fawkes, or indirectly through the study of folk heroes, e.g. Robin Hood. The story was felt to be the most important and the history as much as it was, was the background. (First school history co-ordinator, September 1994)

Contemporary televisual genres such as docudramas, mini-series and soap operas abound with such dangers: 'for our hypercredulous generation . . . there are different levels of truth . . . So what else is new? Helen was hideously traduced by Homer' (Howard, 1992). This amusing analysis of history and fiction's problematic media relationship coincidentally adorned the same page as a contrary article arguing that:

> The culture of the Enlightenment — our culture — has developed conceptions of radical tolerance and liberty which have questioned and undermined the process of its own transmission . . . In the end John Major and John Patten will be judged by their ability to defend the illiberal insistence that this is what and how we must teach, because this is what and how we were made. (Appleyard, 1992)

No convincing evidence was offered to support such a view, and retrospect can help us judge Major's and Patten's successes in imposing the educational absolutism called for. My own experience is that teachers tend to be more self-critical and realistic about teaching methods and curricula. 'Living history' according to one of its leading proponents can lapse into:

> Fun and games . . . as educational as your medieval banquet type caper. And it can be historically very bad — totally misleading and contain all sorts of erroneous elements . . . It's not unique to active history by any means. Look at any textbook and you can find things that even unintentionally, even by the way they are phrased, can create quite the wrong impression . . . (John Fairclough, Interview, April 1993)

This advances the question beyond the binary simplistics of 'seductive historical fiction' versus 'solid and authoritative educational fact': modern children experience many media, and principled and knowing uses of story can educate their citizens' susceptibilities by providing reflexive, efficient and motivating forms of teaching and learning.

In Practice (6)

Stories in National Curricula

National curricula ebb and flow, but stories are always with us. Examples of two approaches are given below, deriving from 1990s national curriculum practice but essentially timeless. The first describes starting points for younger children's story-listening and making. The second is a typically-structured historical story for telling to older pupils. Both examples are premised upon the principles described so far.

Some Storied Starting Points for Younger Children

Most teachers, classroom assistants, visitors, family and friends have a rich bank of storied memories which can introduce children to history:

- a favourite toy or game;
- ride-on toys like a bike or car;
- a memorable scar or accident;
- a first club or gang;
- an early book or television programme;
- being told off at school;
- earning pocket money;
- a first day at work;
- stories of birth and infancy.

How are they made and told? (e.g. photographs, albums, videos, memories, souvenirs, timelines).

Having heard and played with examples such as these, children can start to use story actively to research their own lives and surroundings:

- What are the stories of my personal and family names, my birth day and month, my pets' names, the names of the places where we live and our school? Research builds into a booklet describing 'the story of me'.
- Do I always get along with other children? Stories of quarrels and how older children feel when a new baby arrives.
- What is my favourite treat or day out? What can older people remember about theirs? Class book of special day stories to read and compare.
- How old is the house I live in and what is its story? How do we know when things, people or houses are old?

- What can stories tell us about homes from history? Using photographs and drawings of real or typical houses, what stories can we devise about the people that used to live in them?
- Did Grandma have 'Toys Я Us?' Where did my family buy toys and sweets when they were young? Teachers, families and other helpers tell shopping stories in a class book for reading, telling, illustrating and comparing.
- If we wanted to play 'old shops' how would we know what to do? Do we know anybody who used to work in a shop? What can stories and pictures tell us about old shops? Together we will make an 'old shop' corner just like the real thing . . .

A Storied Starting Point for Older Children

Boudica's War

Discussion about this story is supplied in italicized script below, and a text for the story in roman script.

Background

If you do not feel confident about telling stories 'live', use a mixed approach. Read from a text like this, with intermittent props or activities to make the story real and interactive.

Boudica's story symbolizes some of history's intractable problems. There are no written sources portraying the Boudican revolt from the Celtic point of view, and none strictly contemporary with events. *Boadicea* as a name is a medieval clerk's error and even *Boudicca* with a double 'c' derives from Tacitus's mis-spelling 50 years after the revolt (Webster, 1993). Boudica never rode in chariots with knives attached to the wheels — consider the consequences for her own troops! Most of what we know about the Celts of England is from European, Roman, Irish or Welsh sources. Her tribe the Iceni may have collaborated with the Roman invaders since 54 BC, so it is surprising that Boudica became a role model for English nationalism in Victorian history books.

Share and discuss such different *Boudica* stories with your children. Tell them about the version you are using. It is written from the retrospective viewpoint of a Celtic participant and purports to be plausible: but we have no surviving words from such a person. Offer simplifications of the original evidence, that space precludes from inclusion here. Above all encourage reasoned speculation about as well as enjoyment of the story.

I am a Celt. I am lucky to be alive, but so ashamed to be running away. And it all looked so good just a few weeks ago. We were chasing the Romans for once, instead of them chasing us. Seventeen summers they have been at it, telling us what to do in our own lands. (*Perhaps have replica Roman coins and grain (bread), chains (slavery) lead and iron here to represent why the Romans invaded Britain*).

Boudica stopped all that. What a woman! Her husband the king died. The greedy Romans wanted his money and his kingdom. Boudica said 'no'. The Roman soldiers stripped and whipped the queen and her daughters, but they would not give in.

Hurting our queen started the trouble, but it was not just that. Boudica wanted all the tribes to fight because the Romans were annoying everybody. The Roman army was killing our holy men and women. The Roman traders and tax collectors always wanted more money. The old Roman soldiers in Colchester kept taking land from people who had lived there for years. They built a whole new town. Then they spent our taxes on building a temple to the man who invaded us 17 summers ago. It was an insult. (*That man was Claudius — with other stories to tell of how a lame, stuttering leader brought elephants to Colchester to celebrate victory*).

They are not human, these Romans. They want to rule everything, buy everything, write everything down. We hate them. Boudica and the priests asked the Gods what to do. The Gods chose WAR. Then we showed them how Celts could fight.

First of all, Colchester. The Romans hid in the temple of their dead emperor. We ambushed the army sent to rescue them. Then we burnt the temple, with them still inside. Next, London. The fools could have run away but some of them loved their money and houses too much. We killed them, slowly. Then St Albans. What we did there will teach Celts not to mix with Romans . . . (*Perhaps allude to Celtic barbarities, though these are too strong for many children*).

We walked for days up one of those boring Roman roads. We were looking for their army and burning a few farms. We had 10 times the warriors they had. At last, the Romans stopped to fight. The time had come to decide: who would rule? We were sure we would win: some people even brought their families to watch, in wagons. Boudica had led us all the way and now was our chance to get rid of the Romans forever. The chants, the dances, the war cries began. Our chariots raced up and down. We were wild for a fight . . . (*Perhaps show printed or TV pictures of Celtic warriors and Roman soldiers*).

That was the trouble. We were wild — but the Romans just stood still, quietly. It got more crowded the closer we got to them. Then — whoosh. Thousands of Roman spears. The earth turned red. Whoosh — thousands more. We couldn't move for dead bodies, but the Romans marched on. They were hiding behind their shields, the cowards, then stabbing us with their sharp little swords. That was when the real killing started: men, women, children, horses. If only all the tribes had joined in: we might have won.

I ran and I am still running. Who knows what happened to Boudica? But the Romans will write it down. The Romans write everything down . . . (*Perhaps use simplified documentary sources for further study*).

Chapter 7

Making Tradition Critical?

History, real solemn history, I cannot be interested in . . . the men all so good for nothing and hardly any women at all . . . and yet I often think it odd that it should be so dull, for a great deal of it must be invention. The speeches that are put into the heroes' mouths, their thoughts and designs . . . (Austen, 1818, p.123 — Catherine Morland talking)

When I was at school I hated history — it was only when I started to read historical novels that I grew to like it and now I can't get enough. (Primary school teacher during course, June 1994)

Opening Edmund? A Story in Action

In retrospect and as I write it becomes obvious that the story of Edmund, the last and martyred King of East Anglia, has always been with me. I grew up in his town (Bury St Edmunds) and attended his school (St Edmundsbury Primary 1963–67). One of my powerful childhood memories of history, alongside visits to remote Scottish castles and mystical Cornish standing stones, was a *Son et Lumiere* in Edmund's 'Abbey Gardens'. In the darkness and from a grassy bank I watched as lights and actors played out a story of Viking invaders, river-sailing longboats and Edmund's Christian martyrdom. It was 1966, it was hammish, it was inaccurate and I subsequently took two history degrees. Later, as an advisory teacher, our office overlooked the fishponds and vineyards of the Benedictine abbey founded to venerate Edmund, and NC history offered many opportunities once again to make Edmund do some educational work.

I have consequently told versions of the Edmund legend in numerous schools, structured simultaneously to transmit the story and analyse its provenance: ideally *evidence informing narrative*. In the following example the oral tradition had already been raised by the teacher: 'We'd done a bit on Beowulf you see . . . they know about the idea of handing the story down, how it gets altered. They were quite taken with that idea' (interview with class teacher). Despite this the teacher agreed that in later poetic retellings the children had 'basically kept the version of the story they were told'. Was this teacher-dominated, cosmetic transmission or the start of genuine historical questioning? I leave it for the reader to decide, through cited extracts; evidence from one storied lesson typical of hundreds I have taught.

A rough written plan sketched the order and form of the lesson, and a booklet (Suffolk LEA, 1991a) provided teachers with supporting documentary extracts from contemporary historical sources. On this occasion I chose not to plan predictively for specific history or English learning objectives, preferring to be guided by children's instantaneous reactions, by my own well grounded knowledge of the primary curriculum and the class teacher's descriptions of the children's previous learning. NC documentation and ideas therefore influenced this teaching but were never in total control. The school itself was an urban Roman Catholic 5–11 primary, abutting a predominantly local authority housing estate. The group was 24 eight and nine-year-old children, whose class teacher was also the school's history coordinator. They had been studying the Anglo-Saxons for about three weeks. The lesson lasted a school morning: we started and finished in a classroom and enacted the story in the hall. I returned later to interview children and to read resultant descriptive writing and poetry, omitted from this report due to lack of space.

The present version of the lesson has been edited from a full transcription, although practical constraints prevented individual children being identified. Extracts are headed by questions, to help readers construct their own stories. Direct reporting of the words spoken are in normal type, summaries of dialogue square bracketed and italicized [*like this*]. To monitor the author developing and grappling with Chapter 6's teaching principles and learning outcomes (see pp. 86–8) critical comments are bracketed and emboldened where principles have been contradicted (**e.g. Contra principle 2.1**). Short commentaries also end each extract, pointing towards where principles have been reinforced. Section 2 of this chapter then sketches a brief educational history of the Edmund legend, comparing its place in medieval curricula with contemporary attitudes and experiences.

Extract 1: What Kind of a Knowledge Is History?

Teacher:	As I always say to them Mr Bage it doesn't matter if they are wrong, if they guess, give their ideas . . . as long as . . . you are thinking.
Grant:	[*I used to teach PE and Maths. What are they? They answer*] Now, I just teach history . . . what is history though? [*Children say past, years ago*]
Child:	. . . about lives of people? [*I agree*]
Grant:	Yeah . . . the people that used to be alive . . . I wonder how much they were like us . . . how much they were different? [*This lesson about that. Explain I tape record lessons*] . . . so that I can listen to them later on to see whether I did them right or not . . . sometimes I don't. [*I now challenge them to tell me ten things about the Saxons*] . . . anything at all — it doesn't have to be very clever.
Child:	They had thatched roofs.
Grant:	[*Yes but*] . . . how did you know that by the way?
Child:	Because Miss showed us some pictures [*We look at West Stow postcards*].
Child:	Their houses didn't have windows?
Grant:	[*Good boy, explains further, that's three*].
Child:	They didn't have chimneys, they just had holes in the roof.

Grant:	Yeah, that's right, so it must have been quite smoky . . .
Child:	They came to Britain? [*I reply yes and ask where from. Child replies Germany*].
Child:	Um, they lived, they put, some of them put their houses in the countryside?
Grant:	Yes . . . they didn't have many towns so Ipswich wasn't a great big city in Saxon times. It was called Gippeswic, a tiny little village really. Well done.
Child:	They had a candle at night?
Grant:	[*Interesting because some Saxon people had candles, some didn't*]. The people that did were richer . . . monks who lived in churches or nuns . . .
Child:	They needed land to reward their soldiers?
Grant:	[*Yes, bit like the Romans*].
Child:	They had to pay for people's lives?
Grant:	Yeah! . . . tell us a bit more about that.
Child:	Well, if part of your family was killed, the person who killed them would have to pay some money and if you were working or rich you would pay quite a bit.
Grant:	[*Yes. Demonstrate with another example*] Because the Anglo-Saxons didn't . . . really have proper prisons. (**Contra principles 2.4, 3.7, 4.6 — the word 'proper' is inappropriately judgmental and anachronistic**).
Child:	Some of them came to Britain as pirates.
Grant:	[*Repeats, yes. Romans arguing, Saxons came*] And gradually it came to be that they forget they came from Germany and it seemed like they'd been here for ever. (**Contra principle 2.1 — what evidence is there for this?**) And I'm going to tell you a story about somebody whose home was here . . . Edmund.

Comments

The extract opens with appeals to think and a brief recap of the nature of history. I consciously do not belabour links between history and 'evidence', preferring to examine the idea later in the storied context. As in the next extract, I probe and articulate the children's prior knowledge so as to ground their understandings of the story in prior knowledge — though this led to some dangerous generalizations towards the end.

Extract 2: What Do the Children Know about this Story Already?

Grant:	[*Saxons, pirates, villages, wooden houses, thatched roofs*] That's the sort of land we're talking about and we're going to go back to . . . through our story . . . about Edmund . . . a special person for several different reasons. One . . . he became a king when he was very young. Now how old are you? [*Mostly eight or nine. We discuss 16-year-olds they know*]

Grant: [*All know or can imagine a 16-year-old: any of them kings or queens? Children giggle. Edmund that age was king of East Anglia, Suffolk and Norfolk. Explain Norfolk by example of Norwich*] But there weren't really any big towns . . . any roads like Roman roads with paving stones . . . (**Contra principle 3.7?**) [*No motorways, factories. What sort of jobs? Any checkout girls or boys . . . any nurses? Pairs: two minutes to decide Anglo-Saxon jobs*]

Child: A bootblack?

Grant: What, somebody to polish shoes? Now that's a good job from the past but it goes only about 100 years ago and the Saxons were about 1000 years ago . . . yes, it is Victorian, you're right. Any other jobs? I'll come back to you . . .

Child: A blacksmith? (**Perhaps much of this section contradicts principle 1.4 — were the children 'guessing to please teacher' as well as activating prior knowledge?**)

Grant: Yes, they certainly had blacksmiths . . . [*explains what they do*]

Child: Wood carving?

Grant: Yes, well done [*for bowls, house decoration*]

Child: Pottery making?

Grant: Yes . . . [*pots, cups, jugs, clay diggers*] . . . I'm going back to those boys.

Child: Weavers?

Grant: Oh yes, very clever, good — they had weavers who made cloth by putting threads like that together . . . [*gesticulates*]

Child: Slaves and soldiers? [*yes, especially in this story*]

Child: Village doctors?

Grant: Well they didn't have [*doctors, ambulances, nurses*] . . . but they did have people in the villages who knew a lot about . . . [*childbirth, wounds, temperature*]. But they didn't have many tablets or medicines. Can you think what they did have?

Child: Plants, plants, herbs. [*Yes, back to boys*]

Child: Farming? [*Yes, many, because grew own food*]

Child: Jewellery makers?

Grant: They did. [*Few diamonds, clay, glass*] . . . But the Saxons weren't very good at making glass themselves. (**Contra principles 2.1, 3.7 — this is a generalization extending too far chronologically and geographically**) [*Winemakers, thatchers, spinners, dyers are cited*]

Grant: Yes . . . In a minute you're going to be someone with a job.

Child: People who made weapons?

Grant: Yes! . . . and of course, Edmund being a king would have had special things and special weapons.

Child: A horse man.

Grant: Yes . . . some people had horses if they were quite rich . . .

Child: Tack, did they have tack?

Grant: Tack? For horses? They had leatherworkers [*cups, armour, clothes*] . . . so yes.

Child: Teachers at school?

Grant: Well . . . they didn't have schools like we do but yes, they had people who taught things. [*Storytellers, poets, monks, nuns ran little schools*]

Child:	A taxman?
Grant:	There is always, everywhere, a tax man, isn't there? . . . Every people in history have had taxes . . . **(Contra principles 3.5, 3.6 and 3.7 — another heat of the moment generalization that could have been rephrased)** Very good, they had tax people.
Child:	Cooks? (*yes — one more quickly*)
Child:	Coach driver?
Grant:	They didn't have any coaches . . . a few wagons and carts but . . . It would be about as common nowadays as having a helicopter I reckon . . .

Comments

I start by discussing their knowledge of contemporary 16-year-olds, enabling them to put Edmund into a personal perspective. We continue to work from their historical knowledge by discussing possible Anglo-Saxon jobs and raising anachronisms and misunderstandings. The resultant fictional workers are perhaps stereotypical but will people the story of Edmund; their creation can help the children to enter this imaginative reconstruction from evidence. They communicate this physically through mime in the next section.

Extract 3: How Can I Help Children Enter the Story?

[Miming work is the next task, for others to guess at. I show a 'bad mime': what's wrong? They tell me that I wasn't looking and needed expression. I add 'make it look like hard work'. We move to the hall and view mimes such as: weaving, spinning, sewing, building, blacksmithing, beermaking, herb gathering, carving, farming, cooking, chopping. There are frequent incorrect guesses, questions and answers from child to child, although I channel and comment upon most.]

Grant:	Now . . . go and be doing your job . . . When you hear me clap, I want you to freeze and I'm going to tell you something else . . . FREEZE. Villagers, there is some bad news . . . We've heard in the last few minutes that Vikings have landed. The Vikings, pirates **(Contra principle 3.7 — though there is sound evidence that the Vikings in this particular story were piratical)** a people from over the sea just like the Saxons, us Saxons came from over the sea to settle here in East Anglia. So, another people from Denmark and Norway . . . They have a huge army of between 30 and 40 shiploads . . . All are fighting men, with battle-axes and swords and shields and hate in their hearts. **(Contra principle 3.7 as above and 2.1 — narrative momentum is interfering with historical accuracy)** They have come to take our land, our gold, our children as slaves. What should we do? More to the point, what should we advise our King to do?
Child:	Fight against them.
Grant:	You think we should fight?
Child:	We should go away.

> *Grant*: We should go . . . run away? What and leave them our land? And where
> would we go? Maybe . . .
> *Child*: Fight against them for our land.
> *Grant*: We should fight against them for our land? What think you? [*Pause, I
> switch pace, place and tone*]

Comments

This section is fictional: no evidence suggests that Edmund did consult his people
or that they discussed what to do. As a device to help children to understand actual
historical events and to teach that historical and life outcomes are neither inevitable
nor predictable, I view it as pedagogically justifiable.

Extract 4: How Can this Historical Episode Be Connected to Modern Children's Lives?

> *Grant*: Back NOW — have any of you ever been bullied? [*Children say yes, by
> friend down road, boys at school*]
> *Child*: There's my cousin who comes round she bullies me and she takes my
> toys and things and she says that she will bring them back but she doesn't.
> *Grant*: Now, if you let a bully get away with it I wonder what happens?
> *Child*: They'll just do it again!
> *Grant*: Let's go back! To our village! And, be working . . . Villagers, to the
> green quick! [*I shout*] I have some news for you [*Edmund is to give
> Vikings horses, food, water, shelter — he will not fight!*] . . . Those of you
> who have arms, put them away and hide them . . . [*Pause, change pace
> and tone*]
> Now . . . if you let a bully get away with something . . . [*Grant mimes
> taking sweets, Nintendo*] If you don't stand up to that bully, what happens?
> **(Contra principles 2.1, 3.7, 4.6? Most schools advise children to tell
> others about bullying first; and is 'bullying' an inappropriately ana-
> chronistic term for uneven conflicts between warrior communities?)**
> *Child*: They'll keep on coming back again.
> *Grant*: Now . . . that's just what the Vikings did. They went to York and burnt
> it. They went to Northumbria and they got the two kings AND THEY
> SLAUGHTERED THEM! [*ugh, child in background*] They went to Not-
> tingham and they besieged it . . . And when they got fed up . . . where do
> you think they went next?
> *Child*: To East Anglia, back to East Anglia?
> *Grant*: [*You are right. I ask 'should we fight'? 'Yes' but no weapons. Mime
> making some: blacksmiths, leather workers, carpenters. Ask again: who
> will fight? All say 'yes'*] **(Contra principle 4.7 — have I used the
> story's force to produce a false consensus?)**
> *Grant*: We have a few days . . . Even if you are a farmer . . . or a monk . . . go
> and make weapons . . . and remember it is with spears and shields, swords
> and axes and leather armour that we shall defeat the enemy. But above all

it is with Christ in our hearts . . . [*Children mime and show weapon-making*] (**Contra principle 2.3 — despite 'knowing' intellectually about imbalanced sources — and being brought up as a pagan — I have manufactured righteous feelings of a Christian and militaristic consensus. But would anybody have fought without these?**)

Comments

I have been explaining and moralizing using the historical analogy of the 'Vikings as bullies', conscious of this as an heretical act against the functionalist, detached view of 'history as analysis of evidence'. What disturbs me is whether I have sufficiently underscored my intended buttresses against inculcation. Are the contrasts clear enough between my own views and time, and historical equivalents? Are the 'choices' about whether to fight real? It would be a brave child who refused to! This may be accurate, since the urge to conform is a powerful force in history: but is it ethical? Similarly, although the invocation of Christian faith was recognized by these Church school children, my own sympathies and belief systems lie as much with the Pagan Vikings, whom I have traduced. These questions are unresolved in the next extract.

Extract 5: What Moral, Personal and Historical 'Choices' Might this Story Entail?

Grant: What have you brought for the day of doom . . . ?
Child: Catapults.
Grant: Catapults? . . . What will it throw? [*Replies stones*]
Grant: Stones? Well . . . we are not fighting in castles. We shall meet the Vikings in an open field [*Maybe stones in reserve. Continue: spears, shields, swords, daggers, armour, bows, arrows, axes, knives are all brought through mime*]
Grant: Now, we have a glorious array and spread of weapons . . . But before we go into battle you must know: [*Viking army bigger, hardened, led by Inguar and Ubba. We are farmers but will still fight. Children repeat names of Viking leaders*] Inguar and Ubba are pagans . . . not Christians, as are we . . . they do not believe in Jesus . . . And they — worship — the — god — of — WAR. They will take no prisoners. It will be a fight to the death, a fight for Edmund and for Christ and for East Anglia. Or, death at the hands of the Viking pagans. (**Contra principles and reinforce worries as above**) [*Pause*]

Who now is prepared to stand up and bring their weapons? If you have courage, pick your weapons up and place them in your belts or over your shoulders.

[*Children line up on 'crest of that gentle hill', make shield wall before Edmund. Wall steps forward once, twice. Listen, the Vikings approach chanting*] **Thor, Thor, Thor**, you can hear the sound of their hate as they bang out the rhythms of their gods' names on their shields.

[*If I touch somebody, they 'die'. Step forward, close ranks, repeat twice*] This time the Vikings send forward their best warriors with double headed axes [*child — ugh*] . . . your friends are kissed by the maiden of death and the shield wall closes . . . [*repeated again*]. For death is what came . . . One by one, Edmund's brave army were hacked and speared and sworded to death apart from just a few . . .

Right again warriors, back to the village! And let me tell you, after we have had a break . . . which of you survived, whether Edmund was one of them, and what happened after that. [*Playtime*] (**Contra principle 1.5? There will be discussion at the end but the lesson could end here to allow time to examine meanings**)

Comments

I try to communicate the reality of violence through an imaginary recreation of a documented battle: but there is no actual contact, and the Viking army exists only in our minds. To exclude violence in such an 'heroic' story would dangerously mislead the children about past and present, but the element of 'choice' is minimal for the child actors.

Extract 6: How Do We Know this Story?

Grant: Somebody just after the battle wrote down what happened . . . but only about three sentences. Hardly any writing was done in Saxon times . . . (**Contra principles 2.6 and 3.7? This is the sketchiest of references to documents AND misrepresents Anglo-Saxon interests in literacy**) And this is what we think happened. [*Pause*]
We know that there was a battle . . . the East Anglians were defeated . . . Edmund was not killed, we don't know whether he ran away or escaped . . . But he was chased by the Viking kings. [*Children repeat Inguar and Ubba*] We don't know whether the story is completely true, but this story has been passed down because of a 12-year-old . . . Inguar and Ubba chased Edmund to a nearby wood. Some people say there was a hut or a hall there. They . . . got Edmund in a corner . . . all Edmund's soldiers had been killed or run away . . . ALL apart from one 12-year-old . . . Edmund's armour bearer, someone who carried the weapons. They weren't like a grown up soldier . . .
Now . . . Edmund . . . was surrounded by Viking soldiers. [*Inguar and Ubba in dialogue offer a deal to Edmund. He can stay as king if he obeys them but must also renounce Christ*] . . . and come back to the old gods, the true gods of Thor, and Woden, and Freya . . . the gods that you Saxons used to worship. [*I summarize the deal they are offering*] All this was seen by the 12-year-old armour bearer . . . Maybe Edmund had fought his way out, maybe Edmund had killed many Vikings already, maybe Edmund was wounded? None of these things we know, but we do know that . . . it was because the armour bearer was brave enough to follow

Edmund, hide in the wood and watch what was happening that I can tell you, now, the story. (**Contra principle 2.1, 3.5? As revealed in section 2, the sources for this interpretation are few**)

[*Vikings offered a good deal. What did Edmund do? I survey opinion. Children say give up Jesus. He did not because he believed*]

Grant: Let me tell you what the armour bearer told as a story . . . he says that Edmund was tied to a tree . . . [*Child says 'know it' I say keep it to yourself. Choice repeated: take orders, forego Christ. As a Viking, I ask again*] What will it be Edmund? Edmund said 'Never shall I give up Christ'. They took their arrows [*mimes creak, kirch, ugh . . .*] and an arrow strikes Edmund in the body. And they ask the question again . . . [*repeats mime, Edmund refuses*] And the armour bearer tells us that Edmund finished up bristling like a hedgehog. Edmund was in great pain . . . but still he uttered the name of Jesus 'Jesu, Jesu'. The Vikings grew so angry at his defiance that Inguar or Ubba eventually strode up . . . drew his sword and CHOPPED off his head. [*Children go Ugghhh . . . pause*]

The Vikings left the body of Edmund bristling like a hedgehog with arrows, and they took the head by its hair and as they walked away they threw it into the wood. Now, this much the armour bearer tells us.

Somebody, perhaps one of the villagers who survived, perhaps one of you, perhaps the armour bearer, somebody went to tell what had happened . . . [*People come to see. Children are in the dialogue now. We must bury him and tidy the body. Where's the head? All the children search*] And the villagers . . . searched the wood . . . not like a wood today, because the woods were wild. [*Miles, thousands of trees with deer, boar, other creatures*] They were thick and . . . dark. [*The children-as-villagers join hands, comb the wood*]

As darkness was falling . . . they heard [*voice drops*] 'Ruff, ruff' [*a barking*] . . . it sounded like a dog, no, it's not a dog — 'Oooow, oooow'. It's a **wolf** . . . [*Saxons frightened of wolves: eat goats, sheep*] But here was a wolf, barking in a wood . . . Gradually, the people gathered round [*sound coming from thicket*]. And as they did so . . . as we are gathered round today, and as the final brambles were parted there in the middle of the bushes was a magnificent wolf . . . and between the wolf's paws lay Edmund's head [*Ugh, wolf fades into dark*] (**Contra principle 2.1 — this and events below are the most legendary aspects of the story and yet I am not differentiating them for the children**)

The villagers took the head back . . . put the head with the body and lifted the body into a coffin. And they made a small church . . . It was a place for people to go and remember Edmund and worship their god. After a short time the strangest of things started to happen. [*Mimes blind touching coffin, can see, lame walking*] Miracles, miracles . . . when people with faith touched Edmund's body. So the armour bearer says. And . . . after a few more years, people began to say that Edmund is not just a king, a body, a warrior, Edmund is a **saint** . . . somebody decided 'Wouldn't it be a good idea to build Edmund a new church . . . not of wood but maybe even . . . of stone, that was taller and grander?' That's what they did. [*Do children know town in Suffolk? One says Bury St Edmunds, I explain name*]

Comments

This 'tells' the legend as written by Abbo of Fleury in 986, while simultaneously trying to show how it came into being. Its substance is again partially peopled with fictional talk and events for which there is little direct evidence. The development of 'teaching principles' to ground such storied history has relied upon analyses such as these bringing to awareness their reinforcement and contravention.

Extract 7: Do We Trust the Provenance of this Story?

Grant:	Now this armour bearer, it was probably a boy [*gets volunteer*] . . . had a good memory. [*To volunteer: have you?*] The armour bearer hid and watched . . . then went back and told one or two people . . . the armour bearer grew up . . . and became old as we all do [*frail, white hair*]. But he could still remember just what happened on that day so many years ago . . . everybody wanted to hear the story . . . especially a king, about 50 years later, who came to visit the new church that had been built for St Edmund. [*King, bishops and monks listened*] A few years later the armour bearer . . . died. So, how do we know the story?
Child:	It got written down and they found it?
Grant:	Did **he** write it down? No . . . Who was there when the armour bearer told the story? [*Children say king, monks and bishops*] There was a young monk called Dunstan . . . Dunstan . . . liked it so much that he may have asked the armour bearer to tell the story again to him. Dunstan was captured by the story of Edmund when he was a young man . . . And so, he remembered it. And he practised it, and he thought about it, and he asked other people . . . Dunstan started off as a young monk but finished up as a very important man. [*Volunteer to be rich and famous please*] But of course, like everybody he grew old too [*frail, stick*]. Somebody else came to him and said 'tell me the Edmund story, from the armour bearer'. That person was Abbo, a monk, and Abbo could write . . . Dunstan died . . . but Abbo had written it down . . . in about the year 986 . . . just over a thousand years ago. And that's how we know the story of Edmund . . . [*I reiterate stages*] Now you must decide for yourself which bits of the story you can trust. It could be that some bits were made up . . . Why would anybody make up a story . . . ?
Child:	To make it more exciting? [*Yes, could be*]
Child:	To make it more interesting?
Grant:	Yes . . . What about the monks . . . might they have wanted to make it up?
Children:	Yes, no, yes.
Grant:	Why do you think yes?
Child:	Well I don't think they would lie because they believe in God and everything.
Grant:	They wouldn't lie, we hope. But . . . if you wanted to have a nice rich church and lots of people coming in and giving you things, it would be

quite good to have a few miracles happening, wouldn't it? So, we don't know which bits of the story we can trust completely . . .

I think that the Saxons were better at remembering things than we are because they never wrote anything down, and so they used their memory a lot more than we do . . . But you must make up your own minds . . .

Comments

I am trying to show how evidence, as well as the stories we make of it, is transmitted and sometimes manipulated. The children's responses offer some proof of success, though I could have been more radical and probing in questioning authenticity. Elsewhere in the lesson I question the sources' Christian, monkish bias and could usefully have done so here too. My feeling now is that this essentially Christian retelling should have been matched with one from a pagan Viking perspective — or perhaps that at key points a character could have represented the Viking point of view.

Extract 8: What Questions or Responses Do We Wish to Make to this Story?

[*Back in the classroom I compliment the children on their listening and participation, then gave out a printed A4 sheet with six headings*]:

Edmund's army
Edmund's kingdom
Edmund why did you?
Edmund return
Edmund you gave us
How did you feel Edmund?

[*I ask children to respond with single words or sentences. We go through each box offering examples before children write for a few minutes. I explain these notes will become poems. We review a few — examples below*]

Child: Why did you give yourself up?
Grant: Instead of fighting to the death, maybe? Good question.
Child: Why did you believe in God? [*good question*]
Child: Why did you not, um, why did you not give up God? [*same*]
Child: Why did you change your mind a second time and fight them?
Grant: Now that really is a good question. I wonder if Edmund said, 'Go away' the first time to give him the chance to train an army and get weapons? Nobody knows . . . (**Contra principle 4.7 — why is it that as a teacher I nearly always react to a child's comment rather than letting it stand on its own merit?**)
Child: Why did you run away like your fighters?

> *Grant*: Rather than stay and die? . . . Some people think that Edmund did die on the battlefield but nobody really knows. That's a really intelligent question.
> *Child*: Why did you leave us? [*good question*]
> *Child*: Why were you a king when you were 16?

[*Compliment children again as I wind lesson up. Then talk with teacher, referring to booklet's summary of Abbo's evidence*]

> *Teacher*: This child is in the middle of being statemented but . . . Why was he a king at sixteen? That was her question . . . they can surprise you . . . She asks quite well . . . but she can't get it down . . . People say we haven't got enough time to do all of these things but you can use your history for language work anyway. I know it's not the main aim, but it can help . . .
> *Grant*: There was a lot of content in the story this morning . . . It's now set up so that you can talk about it and explain why . . . Which I think is quicker than working through textbooks.
> *Teacher*: It is . . . all the stuff we've done this morning, yes . . . Each year, you add a little bit more, don't you? . . . You get expert with a theme or using the materials . . . I'm gradually just piecing together these bits . . .
> *Grant*: . . . teacher knowledge is important.
> *Teacher*: It is. It gives you confidence anyway — to tell a story or . . .

Comments

This storied lesson helped children to devise questions, only a tiny fraction of which are quoted here: story can move beyond transmission. Story also develops historical intelligence by probing apparently unintelligible things. Edmund did not behave 'reasonably': but discussing his actions may help children to develop the capacity to do so. Significantly, some of the class teacher's thinking about curriculum design and resourcing is also storied. How can we tell and help children to enter a discipline's stories unless, we ourselves know at least some of them well?

An Educational History of the Story of Edmund

Educational storying has an honourable tradition, as I shall now sketch for this case study. Edmund's legend impacted almost immediately on popular consciousness. The Vikings ritualistically martyred him in 869 but within living memory, Vikings had also minted and circulated 'Edmund' commemorative coins for the conquered kingdom (Blunt, 1969, p.254). The first written account derives from Abbo in 986, who claimed he committed it to parchment on hearing Dunstan tell the story. In turn Dunstan had heard it as a youth from a supposed witness, the 'king's armourbearer'. Historical events were being formalized into *a*, perhaps *several* 'moral interpretations'. Abbo's version a century after Edmund's death is already:

A long way from the story told 'simply' by the armourbearer. It has passed through Dunstan's re-telling, no doubt many times, and has been written up by Abbo in the verbose and rhetorical style which he felt the theme demanded . . . a great deal of what Abbo says is embroidery, but it may always be a matter of opinion just how far he represents the actual facts told by the armourbearer. (Whitelock, 1969, p.219)

My argument is that the story was already educational history both as content and understanding. Abbo recalls in his dedicatory letter to Dunstan how the armour-bearer's words were 'stored up . . . in their entirety in the receptacle of your memory, to be uttered at a later date with honeyed accents to a younger generation' (Abbo, in Hervey, 1907, p.9). Abbo wrote these events down because 'when you, the snows of whose head compel belief, made mention of the still continuing incrumption of the king's body, one of those present anxiously raised the question whether such things were possible' (ibid). In other words, and even though the whole was framed by a Christian faith which I do not share, the historical record as presented by the story was questioned, and the written version completed partly to answer that perennial educational question, 'are such things possible?' This was also comprehensive education and lifelong learning, of a sort. Edmund's death became a popular national story not just because of the vernacular literalization of English history through Alfred's Anglo-Saxon chronicle, but *in spite of it* too:

East Anglia's royal martyr, whose fate seemed at first to be of no concern to Wessex, was to become an English national saint . . . he was canonised by public opinion long before he won an official place in calendars of Saints Days in the 11th century. (Derry, 1987, p.157)

After Abbo's Latin came Aelfric's Anglo-Saxon version, and not even the Normans' conquest and critical reassessment of Anglo-Saxon saints could curtail the story's vigour. Aelfric 'was followed after the conquest by various supplementary biographies and accounts by chroniclers . . . as many as 38 medieval sources still survive' (ibid, p.158). Aelfric's tenth century text is of particular interest for he drew on oral tradition through Abbo's writings for explicitly educational reasons (Needham, 1966). During his career Aelfric wrote a history of the world (Hurt, 1972) and about 80 other homilies for oral delivery on 'holy days and saints' days throughout the year' (Needham, 1966, pp.11–12). Aelfric scribed specifically pedagogic grammars and colloquies in English, translating not only Latin words into English but simplifying meaning. 'He usually refers to his works as translations — though they would not be regarded as such today' (Needham, 1966, p.17). Aelfric himself described the process while introducing his Lives of the Saints of whom Edmund was one (my italics):

I have not been able in this translation always to translate word for word, but *I have taken great pains to render sense for sense*, as I have found it in holy writings, *in such clear and simple language as may profit the hearers* . . . I have shortened the longer narratives . . . in case the fastidious should be bored . . . brevity does not always mar discourse, but often makes it more beautiful. (Needham, 1966, pp.17–18)

Might this be an *educational simplification* rather than a *literal* translation (see Chapter 8)? Aelfric's texts were orally pedagogic using: 'rhythmical style ... clearly essentially an oral style, which depends for its effect upon being heard' (ibid, p.23). Later Saxon priests used storytelling and dramatic dialogues in sermons (Cooling, 1994). Until recently I might have distanced myself from the Christian orthodoxies of such medieval teachers; now I perceive numerous educational parallels between their work and my own. Drama, storied teacher talk and the educational simplification of materials through dialogue have a longer history as teaching methods than my previously naïve, modernist assumptions allowed. Nor were these pedagogies cloistered. The monks of Bury were to the fore in producing and sustaining the Edmund legends (Derry, 1987). It was an important part of a medieval churchman's task not just to sustain the physical fabric and evidence of religious stories (such as relics, tributes or buildings) but also the educational, and the mass educational fabric. Thus the eleventh century Norman bishop Walkelin, successfully to end a drought and to underscore the church's power, had Edmund's body carried in procession around the town: 'So great was the throng of men and women present that a man had his arm crushed.' The crowd stopped at a high place and 'there the bishop preached about the saint and implored him that he should intercede with God for rain' (Rollason, 1989, p.233). The abbot of the monastery at the time likewise used his populist educative skills to promote Edmund's story:

> When a villein called Wulmar claimed to have recovered from a coma through the saint's intervention, Baldwin held a hearing with witnesses to test the claim and, when its truth was established to his satisfaction, had the miracle announced publicly in the church, accompanied by a great ringing of bells. (ibid)

A century later Jocelin of Brakelond related how such communicative abilities were still essential to a good abbot, not just in the populist manner displayed by Baldwin or Walkelin. Describing discussions when the abbot's position became vacant in 1180 he has two monks arguing, during which one says: 'How can an uneducated man preach a sermon, either in chapter or to the people at festivals? ... Heaven forbid that a speechless figurehead should be promoted in St Edmund's church ...' (Greenway and Sayers transl.1989, p.13). Jocelin represents Samson, the eventual victor, as 'well versed in the liberal arts and the scriptures, a cultured man who had been at university, and had been a well-known and highly regarded schoolmaster in his own region' (ibid, p.31). This pedagogic approach was sustained in Samson's interpretation of his new job, with which this modern teacher identifies through italics:

> *He was a good speaker* ... He could read books written in English most elegantly, and *he used to preach to the people in English, but in the Norfolk dialect, for that was where he was born and brought up. He gave orders for a pulpit to be erected,* both to enhance the beauty of the church and also *to allow the congregation to hear the sermons clearly.* (ibid, p.38)

It is not just in communicative abilities that I see continuities between Samson and my own perceptions of pedagogic roles. To promote, or to be seen as promoting justice between rich and poor, Samson also investigated:

> Partly by the study of books and partly by hearing cases, he came to be considered
> a discerning judge . . . His precise mind was admired by everyone. The under-
> sheriff said . . . This abbot is a natural investigator . . . he will blind everyone of us
> with his science. (ibid, p.31)

Samson believed in using spoken investigative reason to uncover truth not just in
law courts, but within the framework of biblical revelation, in scripture. His contemporary and chronicler Jocelin of Brakelond (Scarfe, 1997) did not whitewash
over such dangerous intellectualizing. He criticized Samson to his face because
'even in your present position you still cherish the academic notion that a false
premise leads nowhere, and similar nonsensical ideas' (ibid, p.33). What the text
actually says, translated by Greenway and Sayers as 'academic notion', is 'the
notion of the men of Melun' (1989). Abelard, foremost scholar of the twelfth
century European renaissance, had established himself at Melun in 1103 and Jocelin
identified Samson 80 years later with Abelard's school of cross-examinational,
spoken interrogation (see Chapters 4 and 8).

Medieval teachers about Edmund taught visually as well as orally. Thousands
of pilgrims were educated by seen images of the Edmund legend, for the pre-
reformation shrine was hung with 'various rich hangings and carvings, embroidered
or painted' (James, 1895, p.137). Sixty English parish churches were dedicated to
Edmund (Derry, 1987), wall paintings and carvings of the story survive from as far
afield as Buckinghamshire and Herefordshire and a mosque was converted to a
wall-painted Christian church in Edmund's memory by crusading knights in Egypt
(Matten, 1984). 'The story' was made recognizably multi-media, in medieval equivalents, even if it was not transmitted for all the same reasons that as a modern
teacher I might perceive. Yet just as my modernist liberal tradition might desire,
Edmund's story was subject to critical examination and used for diverse purposes.
Popular tales in the oral tradition seem likely to have preceded Abbo's politicized
version 'answering' scepticism about the incorruptibility of Edmund's century-old
body (Hervey, 1907, pp.9 and 57; also Scarfe, 1969, p.306). The story's status
subtly changed from popular legend to being a piece of evidence bandied in a
religious and political dispute. This continued centuries later with Samson, who
similarly applied the legal principles of evidence and reason to the physical remains
of the legend: Edmund's body. Prompted by a near-disastrous fire, in 1198 Samson
speeded up plans architecturally to elevate the tomb. He also opened the coffin
and examined the contents. Leofstan had last attempted this in the tenth century,
according to Abbo, thereby losing his sanity and dying devoured by worms. This
showed the extent to which Samson would risk his soul, as well as his reputation, to
allow *reason* to prove *faith* by Abelardian questioning. Jocelin related Samson's
words on approaching the coffin, in front of 12 chosen witnesses as the abbey slept:

> 'O glorious martyr . . . do not cast me, a miserable sinner, into perdition for daring
> to touch you; you understand my devotion and purpose.' And he proceeded to
> touch the eyes and the very large and prominent nose, and then he felt the breast
> and the arms, and raising the left hand, he took hold of the Saint's fingers and put
> his finger between them.

Six other brothers were summoned to 'witness these marvels . . . and so that there should be plenty of witnesses, by the will of the Almighty, one of our brothers, John of Diss, who was perching in the vault with the vestry servers, saw everything plainly' (Greenway and Sayers, 1989, p.101; Scarfe, 1997, p.3). Samson's prior knowledge is a moot point. Was this genuine investigation or demonstrative experiment? Modern science teaches replicability as proof and investigation as a means to ascertain it. Nevertheless many of the experiments which schoolchildren still experience as science, or the investigations they do in history, are actually pre-ordained in their outcomes even though we hope that through them children can make knowledge their own. Were Samson's aims that different? Whether investigative or demonstrative, Samson knew the power of dramatic story. I do not believe that John of Diss was 'perched in the vault' by accident. Rather in Samson, I recognize a fellow teacher at work . . .

As with 1990s national curricula and histories (e.g. Ball, 1994; Crawford, 1995; Gillborn, 1997; McKiernan, 1993; Roberts, 1990; Weiner, 1993) the Edmund legend had political as well as educational ends and influences. According to Scarfe (1969, p.303) this was partly nationalistic and stemmed from Alfred's revival of royal and Christian interests: 'It is this revival of both Christianity and English patriotism . . . that explains Edmund's prodigious medieval fame.' This was not the whole story though, with nationalism being diversely inlaid with strands of popular justice and 'liberty' from taxes. In 1014 Sweyn Forkbeard threatened to tax or destroy Bury St Edmunds and was struck dead 'by Edmund' (ibid). William of Malmesbury, writing a century later during the civil wars of the 1130s, held Suffolk up as a peaceful tax haven thanks to Edmund (Hervey, 1907). Eighty years later King John's dispute with the pope about taxation, and the appointment of Langton as Archbishop of Canterbury, resulted in John's excommunication and civil war again. Langton arrived in England in 1213 and made his way to Bury. Here, on St Edmund's Day of 20 November he celebrated the martyr's feast and recorded that, in the dispute with John over taxing the abbey following Samson's death in 1211, the monks should stick together: 'divided we shall be ruled: united we shall find ourselves insuperable'. While in England, Langton also just happened to circulate Henry I's coronation charter, the origin of the Magna Carta to which John assented at Runnymede in 1215 (Suffolk LEA, 1991b). By association, Edmund had secured freedom from taxation again; like Sweyn, John was shortly to die (in 1216). Its significance may not have been lost on Edward I. Despite conquering the Welsh even he, in 1294, seems to have drawn back from taxing 'the liberty' thanks to the potency of the Edmund story (Scarfe, 1969). In other words the legend of Edmund served not only religious and patriotic ends; it seems to have become identified, to an extent, with the political humbling of monarchs tempted to exceed traditional limits. Whether, in any of these purposes, the story was manipulated to the point of fraud by the Benedictine custodians of the wealth-giving narrative is beyond the scope of this enquiry. But when we ask the valid question 'who benefits by the stories in our modern NC histories?' it is instructive to realize that the medieval historical national curricular equivalents — of stories such as Edmund — showed similar diversity. It was buttressed against scepticism by monks like Abbo and

Samson using many pedagogic means (story in sermons, in writing, in festivals, in visual arts). It served a variety of patrons (the Abbey, the people, the businesses that benefited from pilgrims) and, over time, fulfilled a range of purposes (to uphold monastic chastity, to bolster nationalism, to humble taxing monarchs, to entertain, to awe). Rather than the contrasts it is the similarities to the modern curriculum and design process which strike me most: curricula and history still serve multiple purposes. For instance, in 1991 a distinguished historian of education could claim that NC history 'was not written primarily with schools, teachers or children in mind . . . It says little about the delivery of the curriculum or about styles of teaching' (Aldrich, 1991, p.2). In 1995 another critic could write of the NC 'The traditional subject-based curriculum was to be a vehicle through which national cultural and moral values could be defended' (Crawford, 1995, p.438). Yet by 1997 a teacher-turned-politician had gained political office and could rhetorically opine that 'the curriculum is far more than pieces of paper . . . Every teacher has personal and professional contributions to make to the curriculum he or she teaches' (Morris, 1997, p.7). Just as Edmund's legend served multiple and sometimes contradictory medieval ends, so does our modern curriculum, and if I as contemporary author am included in this history, the end uses of both have continued to diversify. The search for a singular and definitive curriculum is a doomed and futile task. For as the next chapter explains, teachers do not deliver curricula: *we translate them.*

In Practice (7)

Telling Stories

Many telling techniques are pedagogically ancient, others have been developed by modern teachers-as-story-tellers:

- recording unanswerable questions — step out of role or return to them later;
- maintaining eye-contact with learners;
- posing questions and enigmas in an open-ended way;
- expressing mood through voice tone, e.g. anger, excitement, boredom, etc;
- stopping at points which leave listeners wanting more;
- arriving at your final plan then taking five steps back;
- only starting with everybody's attention assured;
- building in silences and gaps;
- having space to express the story in movement;
- 'seeing' the story in your own mind beforehand;
- changing accents with characters if you feel confident;
- using simple props e.g. 'I am the king when I hold this sceptre';

- emphasizing characters and actions with gesture;
- speaking quietly whenever possible;

(thanks to John Fines and Fiona Collins, see also Fines and Nichol 1997, pp.182–92, Grainger, 1997, pp.146–60).

The following story has been used with KS1 classes to teach history and literacy. Here it illustrates further techniques.

Ada's Special Day

Context

Ask a KS1 class to help you tell a realistic story about an imaginary local child from over 100 years ago. What do we mean by realistic? Is it the same as true? Spontaneously or working from sources such as books, documents or gravestones, children describe the main characters:

- name (e.g. Ada);
- younger brothers and sisters (e.g. Thomas, Anna);
- home and address (e.g. a 2 bedroom cottage);
- father's and mother's jobs (e.g. a farmworker, launderer).

An Example of a Story

The story starts with a question: **how will today be special for Ada?** Then as teller . . . *You describe the jobs Ada normally does at home (e.g. fetching sticks, piling washing into buckets, playing with younger children). Today is important, not for jobs but* **because it is her first day at school.** *Ada has to remember a Victorian penny, a marble, a hot potato. She wears the special hobnail boots her parents have saved for. Mother smiles from the kitchen and waves goodbye.*

You describe her mile-long walk to school and arrival. Two big rooms, a school mistress Miss Charlotte Wilson, three pupil teachers and a stove to stand near as the register is called. There are children everywhere; the other teacher is ill. Ada's name is missing so she says 'Please miss, I'm the new girl and here's my penny.' 'Wait there for the present by the stove,' Miss replies. A present! Ada is excited . . . (Lee, 1962, p.44)

Ada waits and watches: pupil teachers helping with slates, children copying, mental arithmetic, chanting tables, a reading from the bible. It is lunchtime. In the playground she eats her potato and her marble is taken by Albert — a big boy. Ada comes back in and waits by the stove. The class continues working when suddenly...

Miss Wilson enters crossly, large black punishment book underarm. Why must she get angry again? She reads from it, describing past sins... and now Albert has been bullying at lunchtime... she canes his hands then writes in the book... Empty out your pockets too!... 1, 2, 3 marbles confiscated. School dismissed!

Miss Wilson turns. What are you doing here still? 'Waiting for the present Miss.' 'But I meant **for the time being.** *Oh... well have these marbles to make up for it.' Ada runs home and tells everyone all about it...*

Techniques for Telling

The example of *Ada's Story* illustrates general techniques:

- start with a question or statement to secure audience interest (e.g. why was this a special day in Ada's life?);
- move to represent spaces (e.g. working in kitchen, walking to school, standing by stove);
- use changes in volume, pace or tone of voice to differentiate characters (e.g. Albert the bully has a rough voice, Miss Wilson speaks standard English, etc);
- present a mystery object to be returned to later (e.g. the punishment book);
- present a covered item, box or bag to make the audience wonder about its contents (e.g. the cane);
- use repetitive actions to build pictures (e.g. piling washing) and words to reinforce them (e.g. pop the stopper, boil the copper);
- use characters to summarize the story at key points of decision (e.g. Miss Wilson before the punishment);
- leave characters at an identifiable point to make returns easier (e.g. Ada's mother at the kitchen window);
- present characters as torn two ways (e.g. should Ada stay where she was told by the stove?);
- give characters a chance to state their message (e.g. Ada's mother telling her to look after those boots);
- build suspense by characters seeing things before the audience does (e.g. Albert the bully);
- give characters the chance to explain actions by thinking out loud (e.g. Albert sees a girl playing marbles);

- build in a belief that later proves false (e.g. Ada is upset about losing her marble);
- have people left behind who can be 'told' about events later (e.g. Ada's parents);
- let characters display their consciences in public (e.g. Miss Wilson confessing she had forgotten Ada);
- let characters display displacement activities when stressed (e.g. Ada foot tapping);
- enjoy the historical story yourself, because if you don't nobody else will!

Asking Questions

Similarly, fundamental but universal questions helped structure the story of Edmund summarized in this chapter's Section 1. Like the techniques above these can be applied by readers to anchor their own storytelling in educational principles. During or after telling, raise the following sorts of question:

- What kind of a knowledge is history?
- What do the children know about this story already?
- How can I help children enter the story?
- How can this historical episode be connected to modern children's lives?
- What moral, personal and historical 'choices' might this story entail?
- How do we know this story?
- Do we trust the provenance of this story?
- What questions or responses do we wish to make to this story?

Chapter 8

Translating Stories

Chekhov made a mistake in thinking that if he had had more time he would have written more fully ... The truth is one can get only so much into a story ... One has to leave out what one knows and longs to use. (Katherine Mansfield, 17 January 1922, in Brett, 1989, p.21)

Introduction

Chapter 6 discussed some dangers of stories in history, offering as safeguards examples of learning outcomes and teaching principles. Chapter 7 illustrated how these principles developed from practice and offered a case study of a story's educational function changing over time. This chapter will argue that teaching necessitates the manipulation, simplification and communication of any curriculum — its *translation* as in storytelling. Three important phases are outlined: translating history as a discipline, translating national curricula and authoring lessons.

Translating History as a Discipline

Even a cursory examination of recent educational history finds Bruner's influence ubiquitous, especially the idea that 'any subject can be taught effectively in some intellectually honest form to any child at any stage of development' (Bruner, 1960, p.33). This haunted my 1980s teacher education, inspired the School's History Project (Sylvester, 1994) and influenced many history educators (e.g. Blyth, 1985; Claire, 1996; Cooper, 1992, 1995b; Medley and White, 1991). NC history graphically (DES, July 1990b, p.6) and literally echoed this: 'with sufficient care in selection, historical content from any period can be taught at a level appropriate to the understanding of pupils at any age' (ibid, p.10). The ATs and Key Elements of 1991 and 1995 NC histories reflected Bruner's 1960 assertion that 'intellectual activity anywhere is the same, whether at the frontier of knowledge or in a third grade classroom' (Bruner, 1960, p.14). Cooper argued that NC structures offered 'a framework for the development of Bruner's spiral curriculum in history' (1995b, p.vii). But might NC history have been built from over-simplified or misinterpreted Brunerian ideals? Returning to the original text, significant italicized words follow:

'The task of teaching a subject to a child at any particular age is one of representing the *structure* of that subject in terms of the child's way of viewing things. The task can be thought of as one of *translation*' (Bruner, 1960, p.33).

Bruner emphasized teachers' centrality in selecting and translating historical content and ideas. Perversely, NC history then curtailed teachers' professional autonomy to do so (e.g. Dearing, 1993 pp.26–32; Phillips, 1991). Did this matter when Bruner's process of historical problem-solving was defined by NC ATs or Key Elements? In the early days of NC history I and others did not think so. It seemed clear. *Content* was one thing (the study units), *skills* another (1995 AT and Key Elements). Although the two were mutually reliant, *skills rather than content* threaded educational history to provide for 'continuity and progression' (Cooper, 1995b, pp.vii, 29). I now perceive grave weaknesses in these, my former views. Firstly, they placed great strain on the national curricular definition of historical skills. What if these were wrong, imbalanced, obscure, incomplete or inaccurate? Experience shows they were often misunderstood or ignored (e.g. Crawford, 1998; Ofsted, 1998). Secondly, educational distinctions between gaining knowledge through historical processes and content remain unclear. Bruner himself linked intuitive thinking with knowledge of content: 'The good intuiter may have been born with something special, but his effectiveness rests upon a solid knowledge of the subject, a familiarity that gives intuition something to work with . . .' (Bruner, 1960, pp.56–7). Thirdly, advisory experience showed me (Bage, 1993) that teachers were often intimidated by bureaucratic devices like *level descriptions* or *attainment targets* into concentrating upon teaching skills isolated from meaningful learning. This practice was 'uneconomical in several deep senses' (Bruner, 1960, p.31).

The national curriculum's stark content/skills split derived from political sensitivities discussed in the next section, but was an odd phenomenon since the practice was already well established at GCSE (Lang, 1992). Skills versus content may be a crude and outdated tool with which to conceptualize teaching. Emphasizing knowing processes to the exclusion of knowing content can 'far from being a way of challenging established modes of thought . . . be a kind of voluntary marginalisation, a professional retreat into a rarefied scholasticism' (Lively, 1991). An alternative is to ask children to become historians by listening to and retelling rattling good historical stories, and learning how to take apart and construct their own. For this learners need to acquire investigative skills and understand concepts such as change and evidence; but conceived separately, such processes no more define history education than 'pushing round pedals' describes riding a bike. Nor, even if added back together could they be the thing itself. For when we fragment or particularize knowledge we do not summarize it; we change its nature. Polanyi (1983, pp.35–6) illustrated this by analysing a formal speech into different types of activity:

Voice	covered by principles of phonetics
Words	covered by principles of lexicography
Sentences	covered by principles of grammar
Style	covered by principles of stylistics
Literary composition	covered by principles of literary criticism

This does not and cannot justify simplistically linear models of learning. There is no automatic progression from phonetics to lexicography for they are different kinds of knowledge: 'the operations of a higher level cannot be accounted for by the laws governing its particulars forming the lower level' (ibid). Furthermore, summarizers of international research identified nine major 'knowledges' at work during learning literacy: socio-cultural, constructive, tacit, explicit, content, conceptual, discourse, prior and metacognitive (Alexander Schallert and Hare, 1991). If literacy is predicated on such different and complex types of knowing, teaching and learning frameworks must help children move between them. Might language-rich and well-proven pedagogies such as story be superior in this respect to teaching decontextualized literary or historical skills? The emphasis upon the levels of word, sentence and text in the structure of the educationally revolutionary English National Literacy Project (DfEE, 1997a, 1998) highlights this question's importance. For the craft of the teacher is not just to explain the mechanics of levels of knowledge, but to motivate and inform children's mobility between them.

Polanyi's model raises a further and related question. Is being an historian the same thing to a university researcher, a popular broadcaster, a teacher, a museum curator, an antique dealer, an adolescent in a special school or a 6-year-old in an infant school? The answer is clearly *no*: 'I hasten to disavow this always ridiculous assertion. Children cannot be historians in the same sense as university professors can, nor should they be' (Fines, 1987, pp.107–8; also Harnett, 1993). We therefore arrive at an unexpected destination. Although 1995 NC history plotted five usefully clear Key Elements for planning and assessing children's historical learning, teaching consists of much more than visiting these skills, grid reference to grid reference. History, like literacy, comprises a complexity of knowledges. These can be learnt in lifelong fashion and diverse localities. Teachers in different locations develop different principles to help them translate ideas from other locations for use in their own, as sketched below.

A Sketch Map for Lifelong Historical Learning and Teaching		
Description	*Location*	*Translating Principles*
1. Families and individuals learning from history everyday	Before and after school typically home-based and lifelong	Led by translating values, language and the popular media into social or individual experiences
2. Learning history in a group at school within a National Curriculum	Typically at primary and early secondary schools aged 4–14	Led by storied pedagogy translating evidence, ideas and learning

Description	Location	Translating Principles
3. Learning history as an individually examinable activity	Typically at secondary schools and colleges aged 14–21	Led by individuals translating principles of historical criticism and archival learning
4. Learning history as public scholarship and communication, generally constructed by adults	Typically in the worlds of work (e.g. research, media, publishing, teaching)	Led by translating and publishing history for a variety of audiences following principles of accountable reflexivity
5. Learning history for personal and social enjoyment	Typically in the worlds of leisure (e.g. informal learning, tourism, family history, literature, TV)	Led by translating personal interests and feelings into enquiries about the past

Such a model is not designed to be hierarchical. Histories and historians can inform any location if their ideas are translated according to principles developed by practice within that location. This book is concerned primarily with location 2, now examined through educators translating NC storylines and authoring lesson planning.

Translating National Curricula

Disagreement perhaps signals a healthy democracy. In 1990 Margaret Thatcher told parliament: 'Children should know the great landmarks of British history and they should be taught them at school' (Hansard, 29 March 1990). Almost simultaneously an historian saw in the forthcoming NC 'a new historiographical and methodologically reflexive phase . . . wherein history is seen as a problematic discourse' (Jenkins and Brickley, 1990, p.27). The distance between such positions could hardly have been greater and yet the curriculum script from which both read was designed as *common* and *national*. What storylines allowed such varied readings, and what happened in practice as teachers tried to translate national curricula into local realities?

Storyline 1: The National Curriculum as Comprehensive Entitlement

In the 1980s it was feared that history 'would cease to be a mainstream subject in British education at all' (Gardiner, 1990, p.2). HMI monitoring (DES, 1989) showed poor history in much primary school topic work, with curriculum planning too patchy or localized to provide for continuity and progression in learning. Following the 1988 Education Reform Act, government working parties therefore devised (DES, 1990a, 1990b) and eventually amended (DFE, 1995a) a history curriculum to provide pupils aged 5–16 with a comprehensive entitlement. 'Taken as a whole, school history in England and Wales is varied in quality, quantity and organisation. *All* pupils should receive the best possible teaching in history and much less needs to be left to chance than has recently too often been the case' (DES, 1990a, p.2).

This legislation guaranteed history in schools, resulting in significant and arguably beneficial educational changes. Since these have been rehearsed throughout this book, only a few will be summarized here, as a balance against more critical storylines to follow. For instance, *learners like history.* Older primary pupils were recently described as 'generally very enthusiastic about their work in history' with 'positive attitudes towards a subject which at best fascinates and challenges them' (Ofsted, 1998). A national structure helped teachers *broaden curricula* by legislating for learners to study local as well as distant histories, sometimes achieved through story (e.g. Claire, 1996; Maddern, 1992). Pankanahia (1994, p.73) even argued that within the NC 'an anti-racist thread can be woven into history lessons'. 1995 NC history (DFE) advanced the importance of *concepts and skills* in history through identifying them clearly as Key Elements. Although many primary teachers had difficulties with these (e.g. Ofsted, 1998) most secondary and specialist history teachers, in my experience, found them succinct and useful. They can also be advanced using story (e.g. Cooper, 1995a, Chapter 5) particularly through children learning about interpreting, organizing and communicating history. Interpretations have been identified as promoting children's understanding of 'fact, opinion, fiction' (Scott, 1994, p.20). Such ideas are fundamental to meaningful improvements in *literacy* and *language development.* But what other and less benign storylines were embedded in national curricula?

Storyline 2: Herding Teachers and Schools as Scapegoats Lacking Competitiveness

Callaghan's famous prime ministerial speech of 1976 highlighted issues of educational 'accountability' (Brooks, 1991, pp.30–3) and helped usher in 'an appalling decade for government–teacher relations' (Roberts, 1990, p.71). Evidence points to powerful 'anti-teacher' sentiments in parts of the media and political system during the 1980s and 1990s (e.g. Ball, 1994; Ball and Goodson, 1985; Crawford, 1995; Cunningham, 1992; McCulloch, 1997). Suspicion of the supposedly liberal educational establishment became deeply, almost pathologically rooted in some newspapers:

> Industrialists started to raise alarm bells about whether a country whose education system turned out millions of teenagers unable to read or write properly could remain competitive . . . The English group had decided that children did not have to learn Standard English . . . The history group decided that empathy . . . was more important than the history of Britain. (*Sunday Times*, 13 June 1993, p.11)

Assumptions about competitiveness underlay the Dearing Review of the NC:

> Under-achievement threatens national standards of living . . . A nation that does
> not educate its pupils to competitive international standards will increasingly face
> a future which is founded upon the simpler manufacturing functions that can be
> located wherever labour is cheapest. (Dearing, 1993, pp.5, 15)

Alternative views explained failure differently. Basics went unlearned 'not necessarily because they were not taught, but because pupils were not required to learn them in a context which activated and challenged their intellectual powers in relation to the things that really matter in life' (Elliot, 1991, pp.149–50). Such failures persisted, partly rooted in poverty and social exclusion: 'Urban secondary schools in disadvantaged areas continue to struggle to raise standards. Improvements are fragile' (Ofsted, 1997, p.21). National curricula alone had been unable to improve education but Labour policy-makers and politicians continued to argue that raising education standards was fundamental to future 'national success' (e.g. DfEE, 1997b; Morris, 1997, p.3). The next logical step was central direction of pedagogy: 'It is now time to get to the heart of raising standards — improving the quality of teaching and learning' (DfEE, 1997b, p.11). Attempts to achieve this through a National Literacy Framework and Strategy will be reviewed in the final chapter.

Storyline 3: National Curriculum Details as Political Battleground

During the 1980s and 1990s right-wing pressure groups consistently used the spectre of 'politically motivated teachers' to create moral panics about the history curriculum (Crawford, 1995). This may help explain the overt politicization of curriculum policy-making, a phenomenon epitomized by the example of NC history (e.g. Aldrich, 1991). Given education's persistent significance political meddling was inevitable, for NC history was and still is definitive: 'A major aspect of this struggle concerns the idea of "nation". We are, after all, considering a "national" curriculum' (McKiernan, 1993, p.33). In England and Wales NC history's 1991 Final Order represented contemporary governmental educational attitudes (Weiner, 1993) and how 'Thatcherite education policy was conceived, negotiated and delivered' (Phillips, 1992, p.246). English Conservatism took a popularly authoritarian curricular route: 'The traditional subject-based curriculum was to be a vehicle through which national cultural and moral values could be defended' (Crawford, 1995, p.438). History was foremost and according to one critical thinker employed as 'a reconstitutive moral force and as a celebration of oppression and violence' (Ball, 1994, p.39). In this melee the historical representation of women (e.g. Weiner, 1993) and the interests of ethnic minorities (e.g. Gillborn, 1997) often suffered.

Was 1991 NC history 'a remarkably judicious and shrewd compromise . . . a relatively workable and flexible document' (Phillips, 1992, p.258)? Because the curriculum process was being politically influenced by rightist thinkers (Crawford,

1995; Gillborn, 1997; McKiernan, 1993), the 1991 and 1995 NC histories, produced by educational working parties, may have become wary of narratives as conservatively-inclined. The 1991 AT interpretations of history was enshrined as an answer, linking with:

> The old cliché that who controls the past controls the present and the future . . . since you were dealing with historical stories . . . that had to be counteracted by the recognition in pupil's eyes that that might not be the only way of looking at something . . . Without it history could become the sort of thing that a lot of people were worried about. (History Working Group member Tim Lomas, interview, September 1991)

The national curricular end result was a cautious rather than creative view of story. Educational curriculum designers feared potential conservative bias, whereas conservative political influences were so seduced by ideas of accountability and measurement that they became suspicious of anything appearing imaginative or open-ended.

Storyline 4: The NC Marginalizing Teachers

'Is history teaching only about the transmission of knowledge or is it also about what constitutes historical knowledge? These questions have been at the heart of the struggle for history in a national curriculum' (McKiernan, 1993, p.37). Teachers' voices were not clearly represented in such curriculum debating or policy-making (Aldrich, 1991). Take for example the construction of a ten level, integrated 5–16 assessment system in history. As a practising teacher I never believed this could work (Bage, 1993). More important voices in educational history thought similarly (e.g. Dickinson, 1991; Slater, 1991) while research across five LEA's found 72 per cent of secondary heads of department opposed it commenting 'verbose . . . awful . . . daft! There was a general feeling that the assessment scheme seemed to be too complicated and was impractical' (Phillips, 1992, p.255). It *could* have been even worse, as an architect of the 1990–91 history ATs and SoAs revealed while being interviewed about the design process: 'I think my first draft of all the SoAs was something like 160 or 170 SoAs, but we've ended up with 45'. He knew that 'you've got a fundamental problem in history, that children do not progress in neat steps no matter how good you make them' (Lomas, September 1991 interview). This was confirmed by psychological research into historical learning. 'Children's constructions of adult concepts differ sharply from the logical, mature concepts . . . childish understandings change as the child develops . . . Moreover there is evidence that some constructs which seem to be single, simple operations to adults are not so to most children' (Knight, 1989, p.46).

Governments chose to pursue assessment systems designed for neat accountability rather than individualistic learning, despite Stenhouse's warning a decade before: 'Measurement is . . . more reliable, but it is not valid. There is no escape from judgement . . . the objectives model compromises with teacher weaknesses'

(Stenhouse, 1982, pp.90–1). The government's marginalization of teachers, their emphases upon competitiveness and measurement, their politicization of curriculum content, and their negative predisposition towards state education seemed to me, as a teacher throughout this time, to have marred their curricula. Yet I would also argue that curriculum *design* was worse affected than *implementation.*

Storyline 5: NC History and Flexible Implementation

SCAA's chief executive claimed that the NC 're-asserted the primacy of knowledge and narrative in history' (Tate, 1995, p.4). Personally I doubt whether, at the level of practice, narrative and knowledge ever went away in the first place. Put simply, the architects, technicians and theoreticians of the NC over-emphasized their own influence and under-estimated teachers'. For instance I blanched at some of the bitter prejudices flavouring the NC's cultural vision (e.g. Crawford, 1995; Gillborn, 1997; Pankhania, 1994) but, as somebody grappling with it on a daily basis, it could also seem sweet fudge. Working with many partners to translate the 1991 and 1995 NC history documents through teaching, INSET and writing, I found they were chameleons more than monsters. For instance, they abounded with confusing and 'technocratic jargon' (Weiner, 1993, p.87) and were open to different interpretations according to the predilection of the reader (Knight, 1993). Perhaps luckily, legislation is not action and policy is not practice. Curriculum history demonstrates 'a sharp disjunction between structural or organizational change and process or behavioural change at classroom level' (Frater, 1988, p.35). Fullan (1991) identified reasons for this and a major longitudinal study using Fullan's model showed a few KS1 teachers *complying* with NC requirements but most *incorporating, mediating, retreating* or *resisting* them. One fifth were 'active mediators' and 'a majority welcomed the guidelines and structure provided by the National Curriculum' (Pollard et al., 1994, pp.99–102). On a broad level similar processes continued at KS2 (Croll, 1996). Put simply, pedagogy is not a fashion item. Primary teachers neither uniformly enacted 1960s Plowdenism nor sheepishly followed 1990s government dictates (Galton, 1995). Previous research into attempts to change secondary school history showed that 'it is very difficult indeed to achieve basic changes in classroom pedagogy' (Goodson, 1978, pp.45–6). Recent grounded history curriculum development projects emphasized the central role of the teacher and whole class work (Fines and Nichol, 1997). It is into this space that I forward models of teachers as readers, tellers and therefore *translators* of disciplines, curricula and stories.

Authoring Lessons: Storied Planning

Many clever people have learnt history *despite* rather than *because* of being educated in schools (e.g. Collingwood, 1939). A colleague autobiographically and typically described her history as well-learnt from stories and parents rather than classrooms and teachers. 'Every opportunity was taken to visit museums, houses, castles, churches, etc. My imaginary occupants were built from the historical novels and biographies I read during my teenage years' (First school history coordinator, September 1994). From her perspective, *story* peopled historical places in ways that classroom or textbook-based *curricula* could not. By synthesizing historical transmission and analysis, story similarly solves educational problems for me. My personal history pedagogy has become the construction and deconstruction of real stories about the past, derived from evidence and in answer to questions. This is not to argue for the suppression of problem-solving, analysis or quantification, nor for a monolithic, narrative pedagogy. Rather it is to hope that when we translate (Bruner, 1960) history for children, we preserve two strengths. Firstly, tapping into fundamental, subject-transcending ways of thinking such as paradigmatic or narrative (Bruner, 1986). Secondly, finding ways of making this translation easier for teachers and clearer for learners. When we ask children to act as historians and they reply, 'how do historians act?', and we answer, 'like these key elements in NC history', we over-simplify the nature of history and over-complicate the process of education. For modern western culture is increasingly disputatious about the content, form and value of history as an intellectual discipline (e.g. Jenkins, 1997; Southgate, 1996). Nor are NC histories themselves 'true' since they merely represent bureaucrats' and committees' often noble attempts to standardize a discipline in ways acceptable to schools and politicians. Any public or national history curriculum is therefore constructed from a tangled web: political ideas, technocratic and quantitative visions of accountability, theoretical educational assumptions, functionalist pressures to improve 'skills'. These in turn are translated into lessons by teachers, themselves influenced by conflicting values, pedagogic fashions, professional folklore and personal experience and prejudice (e.g. John, 1994a). Children cannot therefore simply act as historians by following a NC because education and history are complex and shifting disciplines. The consequence is that, theoretically and empirically, the induction of learners into disciplines requires their principled translation by teachers. Returning to the experiential and autobiographical thread of this book, for me the simplest and most confidence-inspiring way to achieve this with the 4–14 age range has been by teaching through story. It is not, though, just about *telling* stories:

Journal, April 1992

Might teaching generally be construed as 'storied' and used in PE, for example, to plan the teaching of 'catching'?

- Typically, a teacher might start by asking children about games in which catching matters [*story prompting*]
- The teacher might then demonstrate common strengths and weaknesses when trying to catch a ball [*storytelling*]

- Can the children say what the teacher is doing right or wrong? [*story interpreting*]
- The class might practise catching in different games [*storymaking*]
- Which mistakes did they make and avoid? [*storytelling/interpreting*]
- What will they remember for next time? [*story recording*]

This illustrates the idea, but what happens if you remove the noun 'story' and leave the verbs associated with it? For instance:

- The class might practise catching in different games [*storymaking*]

Storymaking perhaps — but what story? Perhaps 'an action which throws balls following rules of the game and in line with lesson objectives planned by the teacher and outlined to pupils'. Shorthand expressions of this might be *story*, or the *purposes of the lesson*. The teacher sees an 'actual' state (my 8-year-olds are bad at catching), then imagines an improved state (they can all catch now) and the means to make the improvement (showing them this and doing that). The teacher might ask a colleague or book for help (what were you doing last week, it looked really good?).

I now conceptualize such planning as *authoring*: writing imagined realities as lesson plans then scripting them with thirty learners. Children rely upon teachers' creativity to construct a context within which they can then express their own, and author their learning. Education might be planned as storied whatever the nature or structure of the knowledge and content, because pedagogy is a narrative art. The teacher plots a series of imaginary events, tries them with an audience and sees how their reactions alter the outcomes envisaged — rather like writing this book . . .

The idea of planning for more than simple *telling* in storied teaching and learning arose from reflection upon practice, and was developed by it (Bage, 1995). Mingled with children or teachers telling are other identifiable roles and combinations. Those listed below are looser and more classroom-based than other socio-linguistic (Cortazzi, 1991, 1993) or anthropological (Cortazzi, 1993) analyses and concerned primarily with teacher-development rather than research.

Story interpreting	the meanings of stories are interpreted or opened for learners to interpret.
Story listening	stories are listened to for different purposes such as analysis, pleasure, information-giving.
Storymaking	teachers and/or learners develop new stories and interpretations.
Story prompting	stories are stimulated from others.
Story recording	the recording and celebrating of other's stories.

Such roles derive from analysing classroom talk as in the following example. This lightly-edited dialogue was extracted from the transcript of a lesson with a Y3/4 class at Woodland View Middle School, Norfolk (October 1996). It centred around evaluating a video about the Romans and Celts (EHES, 1996) but we join the lesson at the beginning, when I am questioning the children's prior knowledge. In the weeks preceding my visit they had used history books, replica artefacts, site visits, museum collections, photographs and illustrations, videos, group work — a rich and diverse collection of sources and resources. Their classroom displays, written work and enthusiastic sharing of knowledge testified to high quality teaching and learning. This extract raised two other points. The first was that even quickfire, limited verbal interactions can comprise a story. The whole extract can be read as a mutual telling, although below I concentrate on the other storied categories. The second was that teacher talk and story suffused the children's views: traditional moralistic and storied practice seemed to have been overlaid, but not replaced, by NC-inspired handling of evidence.

Grant:	First . . . finding out what you know already . . . If I said to you 'Romans', what's the word you're going to say to me? . . . Just one word? [**Prompting**]
Boy 1:	FORT (*loudly*) [**Making, Interpreting**]
Grant:	Fort! [**Listening, Recording**] OK . . . good word . . . [**Interpreting**]
Girl 1:	Take over the world. [**Making, Interpreting**]
Grant:	Take over the world? [**Listening, Recording**]
Girl 2:	Greedy [**Making, Interpreting**]
Grant:	Greedy! [**Listening, Recording**] . . . Have we got any proof for that? [**Prompting**] Maybe . . . well, you can decide . . . [**Interpreting**]
Boy 2:	Cross? [**Making, Interpreting**]
Grant:	Cross? [**Listening, Recording**] . . . Why should we put the word 'cross' with Romans? [**Interpreting, Prompting**]
Boy 2:	Cos . . . don't know [**Interpreting**]
Grant:	I want you to be able to say why you've chosen that word. Yes? [**Prompting**]
Girl 3:	I know, because they tried to conquer . . . [**Interpreting, Making**]
Grant:	. . . They tried to conquer, didn't they? [**Recording**] . . . OK, another word? [**Prompting**]
Girl 4:	Interesting [**Making, Interpreting**]
Grant:	Interesting [**Listening, Recording**] . . . because you've been finding out all about them, I can see that (*gestures to classroom displays*) [**Interpreting**]

(*naughty, mean, steal, silver, gold, selfish, hunting dogs are stories suggested by seven other children*)

Grant:	OK [**Interpreting**] . . . What about for the Celts? One word! Just think [**Making, Prompting**]

(*thought, happy, peaceful, nice, clever suggested by five more children*)

Grant:	I don't know if I agree with all of your words [**Making, Interpreting**], but that's OK [**Prompting**] . . .

Girl 5: Sad? [**Making, Interpreting**]

Grant: Sad? [**Listening, Recording**] Well, somebody said they were happy and somebody said they were sad. [**Recording**] Why did you say they were sad? [**Prompting**]

Girl 5: Because the Romans might come over to attack. [**Listening, Making, Interpreting**]

Grant: Aah . . . I'd be frightened . . . That'd be my word for it! [**Interpreting, Making**]

Boy 3: Good [**Making, Interpreting**]

Grant: Why . . . ? [**Prompting**]

Boy 3: Because they didn't want to take over the world and the Romans did. [**Making, Interpreting**]

Grant: That's a very good reason . . . There's someone who's really using his head. [**Interpreting**] I'm not saying you're right, but it's a very good reason. [**Recording, Prompting**]

Boy 4: Scared [**Making, Interpreting**]

Grant: Scared? [**Listening, Recording**] . . . I would be! [**Interpreting, Making**]

Girl 6: Brave? [**Making, Interpreting**]

Grant: Brave? [**Listening, Recording**] Were the Celts brave? [**Prompting**]

Girl 6: Yes [**Making**]

Grant: OK, how do you know they were brave? [**Prompting**]

Girl 6: Because they stuck up for themselves when they fought the Romans. [**Listening, Making, Interpreting**]

Grant: OK, how did you find out that they'd fought the Romans? [**Listening, Recording, Prompting**]

Girl 6: Watched a video . . . [**Listening, Making**]

After the lesson I asked the teacher, Linda Grainger, how the children's morally interpretative approach to the Romans and Celts had come about.

Linda: Because we talked about who we are in Norfolk . . . 2000 years ago, who would we have been . . . Romans or Celts?

Grant: There did seem to be a sense of identity with the Celts . . . they seemed quite biased against the Romans, in a way!

Linda: That's probably me . . . and perhaps the video. [**Zig-Zag**] I think at this age they're very susceptible but also . . . we'd have been part of the Iceni . . . This is Norfolk, that would have been our tribe . . . This has possibly weighted them a little . . . perhaps we'd like to be on her side! . . . We actually talked about how you would feel . . . And the reasons why they came.

Cortazzi (1991, p.75) similarly but more extensively collected 123 primary teachers' narratives suggesting that their practice differed starkly from objectives-led models: 'planning narratives rarely mention learning, understanding, developing skills and attitudes. Yet interest and enjoyment are not only frequently mentioned, but are the major features of 22 evaluations' (see also Bage et al., 1999). Teachers liked to be able to 'teach divergently' (Dadds, 1993, pp.260–1) and deviate from plans for three reasons: 'first, flexibility, second, the need for talk and third, and

most importantly, the need to follow children's enjoyment, excitement and interest
. . . three key elements in primary teachers' cultural conceptions of teaching'
(Cortazzi, 1991, p.72). Lessons are the confluence between the planned national
curriculum, and individual children's experience of it. At such a point many primary teachers respond creatively to children's perceived and immediate needs,
rather than slavishly adhere to prewritten plans. Pollard et al. (1994) found that a
main primary criticism of the unreformed national curriculum was that it inhibited
such creativity. How might this error be avoided?

Existing customs of actual teacher thinking could better inform pedagogy and
curriculum design: perhaps we should conceptualize planning creatively as authoring
or imagining, rather than be submerged by the militaristic metaphors of targets and
bullet points? My intention is not to glorify the irrational or the unreasonable, but
to harmonize theories of pedagogic design with practice. Imaginative and storied
planning models may not conform precisely to official criteria (e.g. Ofsted, 1995,
p.66), yet seem as close as these criteria to what experienced history teachers
actually 'do'. One researcher found that only seven out of 28 identifiably skilled
primary history teachers 'planned with precise targets in mind' and that in assessment 'distinctive history criteria were not mentioned' (Knight, 1991, pp.132, 134).
For secondary school history teachers a 'feeling of autonomy was vital' (John,
1994a, pp.33–4) and their planning was creative:

> Research evidence appears to show quite emphatically that experienced teachers
> engage in a planning process which runs counter to the rational or objectives-based
> framework advocated by most curriculum theorists . . . this model still dominates
> most curriculum texts, teacher education programmes and central planning criteria.
> (ibid, p.39)

Powerful history teaching does not deliver diluted academic history, nor transmit NC generalizations, nor slavishly follow Ofsted procedures. Rather, teachers
translate ideas from these sources, mix them with their own views and transform
them into pedagogic forms appropriate for learning in their educational location. NC
history or any other centralized education initiative sails or founders over the rocks
of a question that all teachers ask, every Monday morning, 'What pedagogic sense
can I make of this?' 'Leadership is necessary, authority inescapable. The problem is
how to design a practicable pattern of teaching which maintains authority, leadership and the responsibility of the teacher, but does not carry the message that such
authority is the warrant of knowledge' (Stenhouse, 1979, p.117). Stenhouse's theoretical but grounded model underpinned the Humanities Curriculum Project (HCP
— Gardner, 1993), the Schools Council History Project (SCHP, e.g. Schools Council,
1969) then the Schools History Project (SHP). With GCSE this made real educational advances: 'the chief examiner was astounded by the . . . heights to which pupils
who were not particularly able academically could aspire' (Shawyer et al., 1988,
p.214). SHP then influenced the richly conceptual (Medley and White, 1991) and
acknowledgedly valuable (Ofsted, 1993) ATs of NC history. Was all this owed to
Bruner? 'Without this strong Brunerian rationale for history and its practical mani-
festation in the work of the Schools History Project between 1972 and 1985 there

would have been no chance of success for the process-based national criteria . . . for GCSE history . . . or NC ATs' (Medley and White, 1991, p.6).

Partly, perhaps, but this over-simplified 1960s Bruner, ignored 1980s Bruner and was 'Whiggish' educational history. It is modernist arrogance to construct pre-1970s educational history as content-only, lacking process or thinking skills. In Edwardian times Keating advocated 'for the secondary stage the critical treatment of documents or other evidence' (1910, p.119). In 1895 one writer castigated parroted history from textbooks 'It is not history at all: it makes no attempt to teach children to think; it neglects the moral, humanizing element in this study.' Meanwhile, another sketched her vision of history pedagogy:

> Important as . . . knowledge is for practical reasons, the training of the imagination and of the emotions . . . is far more important . . . The chief work must be done by the teacher . . . We cannot properly leave the pupil to get up a text book, or even to do some elementary research . . . something of the joy of the ballad-maker in pure storytelling, something of the passion of the poet for noble deeds, something of the strenuous ardour of the statesman . . . these, however feebly the teacher of history must feel. (Burstall, 1895)

Teachers have always had to translate governmental, examboard or school curricula into meaningful and educational learning. The really big question is whether we moderns can live up to previous traditions: the principled and knowing use of stories may help.

In Practice (8)

Storied Teaching Emphasizing Text

Many of these methods will be well-known to educators, and can be adapted to lead with either text or talk. I have tinkered with them through my own practice and borrowed many from colleagues. They represent a fraction of the living, communal tradition that is pedagogy.

1. Talking Evidence Through

Children are introduced to working in groups through specific explanations of roles and tasks, building in a structured way to spoken co-operative competence. These include turn-taking, giving and accepting constructive criticism, hypothesizing, challenging, listening for a purpose, summarizing, reporting back, asking further questions, voting, agreeing to differ. Many of the teaching ideas in this book depend upon such talkative cooperation in pairs, small

---

or large groups. Specifically *historical* groupwork synthesizes co-operation with substantive, conceptual and discursive historical content to produce talk that makes and communicates historical meaning. For instance, using different groupings children can complete these tasks:

- From sets of written evidence and/or pictures, decide on everyday jobs around a Tudor house, medieval farm or Victorian shop. If your group was a family unit or work gang, who would do what? Write a list of jobs imagining you were going away for a week.
- As managers of a Greek or Egyptian museum you have been given pictures of objects that were dug up in your country then taken to Britain. Work out which three to demand back, and how to support your case in a letter or broadcast.
- From a list of causes and effects (e.g. the English Civil War, nineteenth century emigration) devise a drama that will teach an audience this story.
- As handloom weavers about to be made unemployed by steam looms, discuss the ways in which you could protect your incomes. Write pamphlets, persuasive or threatening letters.
- Try to convince the rest of your class that the Aztecs should have been left alone by the Spanish, using evidence and your own arguments.

2. Talking Documents

Working from originals or simplified versions children use an historic document to:

- express as a storyboard;
- turn into a story;
- include as a small part of a wider story;
- question, especially the document's author or people mentioned in it;
- imagine as a newspaper story: what would be its headline?;
- illustrate;
- Discuss possible *origins*, e.g. 'how did this newspaper report come about?'; *endings* 'what might have happened after you read this enclosure notice?' and *motivations* 'tell the story behind this letter'.
- Mime it, sketch it or write about it during narration.

3. Talking Walls

Learners design, describe, explain or justify a narrative classroom display (e.g. From Romans to Vikings or The Story of the Armada). Primary evidence needs to be included such as portraits, pictures of artefacts, documents. Explanatory texts and labels connect these. Questions about the resultant story are asked and recorded.

4. This Was Your Life

There are numerous variations:

- Children research an historical character then perform a version of the TV programme with prepared questions.
- Children become a character and devise a summative, re-trospective life story from heaven or hell which can be talked live, taped, mimed or written.
- Goalen and Hendy (1994, p.160) have ghosts returning from the past to comment on how they have been treated.
- Well-evidenced events are chosen (e.g. murders, industrial disputes, natural or manmade disasters). From different pieces of evidence groups devise short dramas, strung together or compared to analyse the whole story.

5. Simulations/Games

Ready-made examples exist via board games, role-playing fantasies, computers and group work: with older children you can devise your own. Most games have storylines and characters for children to adopt. Children need plentiful support, practice and background information for games to be successful and historically accurate, but the discussions to create and play by their rules can be educationally rich.

6. Storytime

Improvise, tell or read historical fiction to whole classes:

- Discuss enjoyment, authenticity, typicality, plausibility, style, etc.
- Children devise their own imaginary historical descriptions from evidence (e.g. a poor Elizabethan's house).

- Contribute individual descriptions of historical events, objects or people to group stories. These are circulated, criticized and then used by individuals or groups to devise and tell a story.
- Children devise stories to explain things, e.g. why did people destroy threshing or weaving machines?
- Children imagine conversations between historical characters about great or everyday events (e.g. the first aeroplane flight, a wet Victorian washday).

7. Poetry Past

Children and teachers:

- Read out loud historical poetry, drama or lyrics (e.g. Tudor poetry). Emphasize that spoken not written versions would have been more familiar then. Do we have oral equivalents?
- Assume the role of the audience at the time of the original telling, and imagine how that audience might have responded.
- Compose and perform in historical style, e.g. a Tudor description, an Anglo-Saxon riddle or poem.
- Write and perform poetry that describes, retells or captures the atmosphere of an historical event or place.

8. News Reports/Reportage (Neelands 1990)

By radio, TV, newspaper or website and an extension of the above. Children cooperatively devise a presentational end-product and/ or conduct live interviews concerning an historical event or development (e.g. the execution of King Louis in 1792, the coming of railways). 'On the spot' or 'eyewitness' accounts and interviews work particularly well. Analyse afterwards: were the questions fair and balanced? How do our stories compare across groups? What would we think if we only had that report by which to judge? How would we change the report if new evidence came in? How do our reports compare to actual ones from the time?

9. Character in Role

Teacher or child researches an historical character and is then interviewed by others in role. Costume and props are not essential. Importantly, the character has the option to 'step out of role' and analyse, criticize or side-step a question before going back into role. Questions and uncertainties need recording and discussing. Vast amounts of information can be transmitted, as well as enquiries framed and refined. Able writers can produce a written document summarizing, then discussing the characterization.

10. Hot Seating

Teacher or child becomes expert on an historical theme or event before being put into the hot seat and quizzed, in or out of 'role'. Discussing, modelling and preparing typically historical questions improves both dialogue and the research that sustains it (e.g. highlight questions like: 'why didn't they?' 'how do you think you compare to?', etc.)

11. Witnesses

Following research someone pretends to eye-witness an historical event (e.g. the 1953 floods) and are cross-examined. This can lead to a dramatic reconstruction. Key questions, the answers offered and historical likelihood and evidence are discussed later, out of role.

12. Trials and Punishments

Choose a character or event. Calling witnesses, cross-examining them, making judgments and deciding punishments stimulates a wide range of imaginative talk. With whole classes and famous cases (e.g. Guy Fawkes) this can be time consuming. A simpler version takes actual records of more everyday crimes as starting points (e.g. theft, assault) and assigns simple roles (e.g. accuser, accused, witness, jury). Try deciding how to punish an Edwardian schoolchild for stealing, an Elizabethan Catholic for publicly celebrating mass, a Roman slave for escaping, then compare to actual examples.

13. What Would You Do?

Individuals, pairs or groups are given outlines of *typical* historical problems (e.g. enclosing a common, building a castle) or *actual* problems (what to do with Mary Queen of Scots, or about the Spanish Armada). Their task is to advise decision-takers on the range of possible actions, and to order them preferentially. This is extremely powerful if evidence for an actual event is examined, as a consequence of which real people made and suffered decisions. Goalen and Hendy (1994, p.160) describe an interesting extension: a character has to walk through a 'conscience alley' made up of two groups offering different and perhaps conflicting advice.

14. Storyboarding, Storymapping and Illustrating

Ask children to turn prose descriptions of events or lives, from textbooks and documents, into sequences of pictures. Storyboarding does this by imagining the stories as a series of scenes, sketched in rough form with matchstick figures and simple labels. Storymapping similarly uses pictures and arrows to explain relationships between the elements of a story, but without necessarily boxing them in chronological order. Illustrating can vary from children retelling a heard story by drawing in a number of boxes (e.g. SCAA, 1991, KS1 assessment materials) through to complex series of cartoons, murals or tapestries. All these techniques help learners to translate a story from one genre into another.

15. Riddles and Games

Use familiar word-games or activities but with historical purposes, e.g.:

- Riddles, fables and puzzles were common in pre-industrial cultures. Try originals out on children (e.g. Crossley-Holland, 1982) then they devise their own keeping answers and language true to the period.
- Instead of 'animal, vegetable or mineral' ask 10 questions to guess an historical character, event or object from a period.
- As a teacher, write short stories or descriptions with anachronistic elements. Can children spot these or write their own?
- Divide texts into sentences or paragraphs, which children re-order, tell and then compare with the original. Can they sum up each sentence in a word or each paragraph in a sentence to produce their own precis or storyboard?

Chapter 9

Story Across the Curriculum?

Introduction

So far in this book story has been forwarded as a way of stimulating educational reflection (Chapter 1), a first principle of human society and culture (Chapter 2) and a flexible classroom tool (Chapter 3). Oral stories in particular have been sketched with living and deeply rooted pedagogic traditions (Chapters 4, 5 and 7) which can lead to educationally principled translations of cultural and curriculum content (Chapters 6 and 8). This concluding chapter has been set more firmly in the period of publication. It examines story linking subjects such as history, science and English, and mediating initiatives such as the English National Literacy Project (e.g. DfEE, 1998) or KS3 national curricula. With this in mind I now highlight three ideas particularly relevant to this claim.

The first concerns *curricula and children.* Perhaps the majority of industrialized countries now have national curricula (e.g. SCAA, 1997) and it is difficult to imagine England reverting to the localized systems of curriculum design prevailing before the 1988 Education Reform Act. A national curriculum is designed with the nation's peoples, systems, values, cultures and interest groups in mind, with compromises brokered between national constituencies. What this can never be, on paper, is a *child's* curriculum. Translating national into children's curricula is something that all teachers do and the best teachers do brilliantly. Since time immemorial, story has played the same transformational games; stories can make national curricula childlike and child-sized, an essential prerequisite to learning. This would entail more devolution of curriculum power to teachers than is currently fashionable and brings us to the second theme of *curricula and teachers.* Given access to taxpayers' purses, it is politically easy to construct national curricular initiatives, policies, plans, committees, targets and task forces. It is more difficult to make these work locally and in practice. This is partly because however many threats, insults, promises or rewards are directed at a system of education, politicians and bureaucrats educate nobody; they rely upon teachers to do that. It is also because learning is a long-term activity needing daily care, but lifelong planning. Even during a period of political stability a schoolchild's curriculum will be learnt under at least three different governments and a huge range of political policies and prejudices; yet teachers, parents, friends, communities and the media teach that child hourly, daily, weekly, yearly. The stories they bring and make for a child's education are of equal if not greater importance than any national curriculum or political policy. Teachers cannot and should not be unthinking slaves to any system, either of their own or a government's making. In this light translating national curricula into teaching stories is a creative and powerful tool with which to resist educational prejudice, or tyranny.

The last idea concerns *curriculum efficiency*. As William Morris (1882) advised, 'Have nothing in your houses that you do not know to be useful, or believe to be beautiful.' This principle should perhaps apply to curricula, which are often dull and ugly affairs. In contrast stories aspire to beauty and utility: traditional stories are often high status cultural goods because, like a well-made ancient step, their edges have been smoothed over centuries of use. It is in this long-term spirit of beauty and utility that curricula should be constructed and it is toward this spirit that stories can assist. Stories can also efficiently bear real educational weight. Because a fine story consists of meaningful words, its telling or reading promotes literacy. If set in the past or in a place, it will lead to understanding more about history and geography. If concerned with material change, it will lead to thinking about science, design and technology. Being mostly about people and their struggles a powerful story raises questions of personal, social and moral education. No teacher would simultaneously draw upon all these themes, but the ways in which stories accumulate cultural capital means that any efficient curriculum should consider them as a central investment, not a peripheral luxury. A classic such story is that of John Harrison, a largely self-educated English craftsman who revolutionized eighteenth century navigational and chronological science by mechanically measuring longitude. As well as leading an internationally best-selling book (Sobel, 1995) his story is educationally telling. It will be briefly examined after a short discussion of the relationships between educational science and story.

Translating Science Through Stories

Children's early encounters with school science often draw upon real or imaginary stories. As a 6-year-old girl recently commented while discussing Channel 4 TV's acclaimed early years science series *Four Ways Farm*: 'I like the stories and then my teacher makes them science.' Here the distinction between fact and fiction, between the scientific and literary, is not so much blurred as irrelevant. The scientific role play below followed similar routes:

Teacher: How do you make the kite fly?
Child A: We run into the wind.
Child B: No, away from the wind.
Teacher: Which is it then? Into or away from the wind?
Child C: I've flown a kite before.
Teacher: You're just the animal we need. How do we fly that kite?
Child C: You have to run like that . . . [*she demonstrates by running around the group holding her right hand up to indicate the string from the kite*] . . . with the wind behind you to make the kite go up. (Hendy, 1996, p.195)

Turning the informal, imaginative teaching and learning characterized above into familiarity with a national science curriculum is difficult partly because, like the author, many KS2 teachers have limited scientific knowledge (e.g. Aubrey, 1995; Harlen, 1996; Kruger et al., 1991; Mant and Summers, 1993). Lacking scientific confidence, significant numbers use classroom strategies which avoid exploring scientific concepts (Harlen, 1996, pp.9–10); others cope by emphasizing closed and dull pedagogy which may 'not only fail children but undermine the teacher's confidence and authority' (Osborn and Simon, 1996, p.16). Science needs to be more accessible to teachers, as well as children. Using story can help because it offers: 'Security to the teacher . . . who feels ill at ease with science. In science the teacher is required to encourage children to raise questions. This can seem threatening unless the teacher already has some idea of the investigations that might arise. Story can provide a context for this' (Hislam and Jarvis, 1992, p.39).

Narrating entails techniques of 'questioning, prediction and hypothesizing — skills more typically associated with science' (ibid). Working collaboratively with teachers these authors developed four storied strategies for learning science: developing thinking skills through oral story, using objects to develop scientific story, investigating within the context of story and employing books as a stimulus for technology and science. Such strategies mirror many previously described in this book, for history. They capitalize upon primary teachers' generally greater ease with pedagogy than disciplines. This contrasts with deficit models emphasizing what teachers do not yet know about science rather than what they do already know about teaching. Such storied strategies are not just, though, a matter of practical pedagogy and curriculum efficiency. The history of science is 'full' of scientists using narrative and metaphoric thinking to assist their science-making:

> Crutches to help us get up the abstract mountain. Once up, we throw them away (even hide them) in favour of a formal, logically consistent theory that (with luck) can be stated in mathematical or near-mathematical terms . . . The metaphors that aided in this achievement are usually forgotten . . . made not part of science but part of the history of science. (Bruner, 1986, p.48)

This runs counter to 'the traditional experience of most people in their secondary education — that science merely demonstrates logically that which is already known' (Hislam and Jarvis, 1992, p.39). To advocate using narrative as part of school science is therefore not an attempt to belittle the importance of scientific experiment, observation, recording and analysis. Stories of science cannot replace fair testing: but can help learners understand why it is interesting and important. As this experienced, science-trained primary teacher from an inner-city school understood, children's motivation matters more than early specialization:

> It's a completely reductionist attitude that if you teach science, the younger you are, the better scientists you're going to have. I don't think that is how people get inspired or develop in a career. I don't think teaching science early is going to give the country more scientists. I think having children who are engaged in their work with imaginative and enquiring minds is going to give you more scientists. (Wendy in Woods and Jeffrey, 1996, p.128)

A story showing people grappling with scientific ideas models and contextualizes such enquiry in powerful ways. *Longitude* (Sobel, 1995) for instance, has scientific measurement and human strength and frailty at its heart. It centres around John Harrison, a rural craftsmen who built his first pendulum clock as a teenager, almost entirely from wood and with no apparent training (ibid, p.64). Harrison was born in 1693 and grew up close to the English shore of the North Sea. He would have known from experience how dangerous marine travel was in an age when longitude could not be measured and therefore position-finding was inexact. By 1714 a £20,000 prize to solve this problem was being offered by Parliament, a reward which lured Harrison throughout his long life and which, in legal terms, he never won. In technological terms, though, Harrison changed the world. His most famous and ground-breaking clocks worked well enough at sea to allow the accurate simultaneous measurement of time in two places, and therefore tell a navigator's distance east or west. Today these large machines rest at Greenwich Maritime museum, more or less on the meridian which they helped to create. They also adorn versions of the Harrison story being used in Greenwich as part of the teaching-as-storytelling project (TASTE). This initiative is based at Cambridge University School of Education. It takes professional storytellers into schools to show how story can improve teaching and learning, and researches and tells 'real stories' from a locality linking that place with the globe. Fiona Collins, the research associate for Greenwich, translated the Harrison story for Greenwich primary and secondary teachers to use for different ends: science, technology, English, history and geography to name but a few. But the most telling point remains that *here is a story worth knowing.* It scores National Curriculum points in Science and elsewhere, yet adds up to more than a sum total of those.

Similarly research suggests that narrative-based, personified curriculum texts can powerfully aid children's motivation. For example, an evaluation of storied comic books and TV programmes teaching science in South Africa: 'Showed the great popularity of the comics with pupils who ... even stole them in order to read them, which they did without difficulty. In the text, a narrative structure is used and the "voice" is the voice of the pupils themselves, using vernacular terms' (Peacock, 1996, p.10). Most teachers in this study did not deploy the texts as their designers had instructed: the materials influenced children immediately but teachers needed more support. Curriculum improvement separate from teacher improvement therefore seems a risky strategy, as an in-depth study of Y7 Canadian Science teachers argued. Bruce and Dorothy's stories showed teachers translating scientific content knowledge by weaving pedagogy as

Threads linking language, student culture, student futures, the curriculum ... and the teachers' personal stories ... These narratively derived understandings have consequences for how we imagine school improvement in science. For instance ... the possibility for external agents ... to bring about change without exploring and working with the teachers' narratives is severely limited. (Connelly and Clandinin, 1986, pp.309–10)

Conceiving of teachers in this light raises two perhaps universal questions: 'what does sustainable educational improvement look like?' and 'how can we best bring it about in schools?' Such issues also pervade how the National Literacy Strategy has been introduced in England, and how English is taught through national curricula.

Translating National Curricula and the National Literacy Strategy Through Stories

Diverse Realities, Diverse Texts

The National Literacy Framework (DfEE, 1998) offers a more standardized vision of pedagogy than the first English national curriculum ever attempted; yet modern children's literacy worlds seem increasingly variable and diverse. In their leisure time many children consume commercial and multi-media cultural products, making teachers insecure about the attractiveness of print-literacy: 'In olden days children coming to school might have had new worlds opened up to them which they had no way of coming upon otherwise: great stories, a knowledge of other fascinating times and places, things to thrill and dream of. Nowadays, all the magic seems to lie firmly outside . . .' (Pompe, 1996, pp.94–5).

Long before and beyond school attendance family expectations influence literacy learning (Brooks et al., 1996), and television viewing suffuses children's literacy experiences (e.g. Robinson, 1997). Some English Islamic children juggle learning literacy in Arabic through rote learning of adult and sacred texts, and in English through 'conventional' western approaches at school (Gregory, 1994). It is dangerous to assume that standardization is the best route to improvement, either in the England or the Asia in which this teacher has practised. 'We . . . need to understand the pedagogy of a culture . . . Stories in which readers can find themselves, stories in which the cultural settings are familiar and relevant, are essential if children are going to find reading interesting and meaningful' (Smith, 1996, pp.185–6).

However much commercial and government-sponsored literacy schemes can assist this process, they still rely upon teachers to mediate them. In this spirit, an examination follows of how teachers can translate the intentions of the National Literacy Project Framework (DfEE, 1998) and national curricula, using stories and multiple texts from history. Many such texts will be written (e.g. reference books, diaries, newspapers) or spoken (e.g. interviews, conversations, songs). Others will be visual (e.g. posters, films, tapestries) or three-dimensional (e.g. artefacts, buildings, sculptures, historical sites and landscapes) relying on languages other than the semantic. This book argues that all can be viewed through narrative spectacles. Nineteenth-century English miners or Irish farmers may not have considered their interactions with landscape as narratives, but their human agency exploited the earth over time to make products full of meaning. In these actions, not just in stories of them, lay story's essential elements: 'plight, character, and conscious-ness' (Bruner, 1986, p.21). Retrospective readings of change also result in narrative elements, if not in a full-blown story. Thus the history of a wood may lie in evidence of its location, geology, topology and changing ownership, shape and names. These may not have been conceived by past people as storied: but a modern human's understanding of landscape will often be achieved through describing its history and natural history as though it were a story (e.g. Rackham, 1986). Such real stories answer the question *how did this place come to be what it is?* In this way so-called non-fiction texts can be as storied as fictional ones, even though they may be overlaid by non-narrative reporting. Non-fictional and fictional literacy, cultural and political literacy, personal and public literacies all rely upon learners experiencing a range of texts — and therefore a range of stories: understanding places, films, talk, icons and images, as well as decoding phonemes and words. It remains to be seen whether the standardized structures of the NLS (DfEE, 1998) will accommodate the different needs of diverse children in pursuit of the ever-widening range of modern education's aims: story's flexibility and ubiquity may assist it to do so.

Planning for Historical Texts Within the NLS Framework

The NLS Framework (DfEE, 1998) is structured around detailed descriptions of teaching content and objectives at three levels: word, sentence and text. It theorizes that if these are systematically taught as laid out, children's literacy will improve. There are multiple premises on which this assumption has been made, with one of the less controversial being that children can be motivated into learning skills at word and sentence level through experiencing texts. History offers a plentiful range of these. As a discipline it is centrally concerned with reading evidence for informa-tion and understanding. When taught well it is also intrinsically interesting. Rich materials and motivating questions therefore put history in a powerful position to improve reading comprehension and writing composition across fictional and poetic and non-fiction texts. History straddles both types. By definition it concerns things that happened in the past: their actors and contexts have often vanished, requiring those learning history imaginatively to ask questions, analyse evidence, reconstruct realities and tell the resultant story. In these senses history is fictional and poetic. Yet most events historians study really did happen and are as empiric-ally demonstrable as the laws of gravity or the value of pi. Historians (mostly) aspire to be analytically accurate, to reveal their sources to others and to write 'true' accounts: to be reliable non-fiction. History offers such rich literacy opportun-ities perhaps because it inhabits these two worlds of information and imagination. When reading and writing are allied with speaking and listening, as this book envisages, history and literacy can be harnessed for their mutual advance. The next two sub-sections illustrate ways in which this can happen:

A. Using storied non-fiction texts in the NLS Framework

Four of the Framework's genres (see also Wray and Lewis, 1997) rely heavily upon the telling of stories about particular events, places or processes:

Genre	Description	Speaking and listening	Reading and writing
Recount	To retell events to inform or entertain	e.g. conducting an interview about a life story	e.g. aspects of another's or own autobiography
Report	To describe the way things are	e.g. describing an historical artefact or picture that only the talker can see	e.g. a newspaper story as an historical eye-witness
Procedural	To describe instructions or sequences of steps for doing something	e.g. after a visit, on which aspects another visitor should look out for	e.g. a letter or list instructing an apprentice about an everyday task
Explanation	To explain processes or how something works	e.g. presenting a radio programme explaining Ancient Egyptian mummification	e.g. an information text explaining the workings of a medieval castle

Two other genres are more advanced. Rather than merely telling a story, they require learners to handle competing or even contradictory stories.

Genre	Description	Speaking and listening	Reading and writing
Persuasion	To promote a point of view or argument	e.g. debating whether November 5th should be celebrated	e.g. letters and petitions about a 'new' canal or turnpike
Discussion	To present alternative arguments and reach a conclusion	e.g. discussing whether the Industrial Revolution improved local people's lives	e.g. biographies about controversial historical figures such as Mary Seacole or Mary I

B. Using storied fiction texts in the NLS Framework

The NLS Framework (DfEE, 1998) divides fictional texts into four main genres, all of which offer possibilities for history and story.

Genre 1. Poetry and Rhyme (traditional and contemporary)

Speaking and Listening and then Reading and Writing

observational/sensual poems;	(e.g. sounds, sight and touch of a graveyard, describing the contents of a museum case or display);
shape poems;	(e.g. an Egyptian pyramid or sarcophagus, Victorian railway);
oral/patterned language;	(e.g. nursery, counting, skipping rhymes, work and sea shanties);
performance/choral poetry;	(e.g. Ancient Greek Homeric poems or dramas);
comparing short poems in a range of forms;	(e.g. riddles, charms, spells, epitaphs, songs, proverbs);
narrative poems.	(e.g. Anglo-Saxon epic poetry, or Baker 1988 — English History in Verse).

Genre 2. Traditional Stories (local and world)

Speaking and Listening and then Reading and Writing

fairy stories;	(e.g. their collection in modern times, their generally pre-industrial settings, their archetypal characters and events);
myths;	(e.g. Egyptian, Aztec, Celtic, Greek);
legends;	(e.g. aspects of King Arthur, Robin Hood, local stories);
fables;	(e.g. Aesop, traditional and cautionary tales);
parables;	(e.g. the Bible, Victorian retellings);
modern adaptations on film or TV.	(e.g. Hercules, Pocahontas, Robin Hood).

Genre 3. Contemporary Stories (local and world)

Speaking and Listening and then Reading and Writing

familiar settings;	(e.g. an old school, derelict building, castle or ruin);
fantasy settings;	(e.g. Middle Earth, Camelot);
extended stories;	(e.g. about an artefact, document, memory);
humorous stories;	(e.g. alternative tellings of traditional tales — Grainger, 1997, p.175);
adventure/mystery stories;	(e.g. modern retellings of the Odyssey, Beowulf);
historical stories;	(e.g. by authors such as Garfield, Lively, Morpurgo);
science fiction;	(e.g. how the future will view the past, technological change, fantasies about time travel);
issue-led stories.	(e.g. slavery, refugees, war, protest).

Genre 4. Plays and Playscripts

Speaking and Listening and then Reading and Writing

performance or reading to	(e.g. Victorian music hall, World War II evacuation);
illustrate an historical period,	(e.g. written by children after studying history to
person or place being studied.	show what they have learnt).

Storied Texts, Literacy and Historical Thinking: Year-by-Year Examples

Many primary and secondary schemes or textbooks offer stories in English or history. My own experience is that at best these need adapting and at worst they are a waste of money. Since they rarely fit the local pedagogic problem, commercial or governmental materials need translating or replacing by home-produced materials. This section offers examples of translating the National Literacy framework (years 1–6) and existing secondary curricula (years 7–9). It accords with the National Literacy Strategy groundrules that 'links with the rest of the curriculum are funda-mental to effective literacy teaching' and that during the literacy hour 'pupils must be working on texts' (DfEE, 1998, p.13). *Literacy* may happen where texts meet skilful teacher and enthusiastic learner, but *history* lies there, too. Personal names, family names, street names, town names and birth dates are not just isolated words helping fulfil the reading and writing requirements for a National Literacy child's reception year (DfEE, 1998, pp.18–19); they can also be joined by teacher or child to construct an authentic historical text, telling a child her life story. The same holds true for the following curricular examples. Making storied texts from histor-ical materials offers children chances to make sense not just of literacy and local stories, but to grapple with the real National Curriculum — their own and other nations' cultures. For instance in the *NLS Framework*:

Y1 term 2 requires children to experience 'traditional stories and rhymes . . . playground chants, action verses and rhymes . . . information books . . . non-chronological reports' (DfEE, 1998, p.22). This could mean the teacher merely reading examples from a big book and children 'doing' teacher-devised tasks in groupwork. Or, it could entail children listening to tapes, watching videos, reading books, observing in playgrounds and talking to older people to find out about stories and rhymes, past and present. Having been supported in their reading of such sources children will want to try out their findings, perhaps in performance. They can also recount them in drawing and writing by devising their own book or display. Such language work can be soaked in authentic historical questions. What stories and rhymes did past children enjoy? What is the oldest rhyme or poem we found? Are today's stories and rhymes the same as when parents or grandparents were six?

Y2 term 3 requires children to experience 'information texts including non-chronological reports' (DfEE, 1998, p.30). An information text compiled *with* and *by* children can start with names from a local graveyard, school log book, memorial or street. It will ask fundamental and perennial historical questions. How old is this name? Why and how did this name get to be here? What and how can we find out about this past life? Graveyards are particularly evocative and interesting places offering opportunities for graphical work rubbing, drawing and copying inscriptions; for numerical and ICT recording; for deciphering, plotting and mapping. At their most basic these are lists but they can be analysed into narratives, for behind each name lies a story. Researching such stories takes children into using books and other texts including newspapers, documents, memories and photographs. They and/or the teacher can then combine fiction and non-fiction by devising or reading 'extended stories' with similar sorts of 'real' or 'typical' places in them such as churches, schools, farms, etc.

Y3 term 2 requires children to experience 'myths, legends, fables, parables, traditional stories, oral and performance poetry from different cultures' (DfEE, 1998, p.34). Studying Ancient Greece offers these through the mythology of the Greek Gods, legends relating to actual places or events, the telling of Aesop's fables and hearing or seeing poetic versions of myths and legends — as with children's versions of *The Iliad* or *The Odyssey*. It is not just, though, a question of this topic offering general opportunities for history and literacy through stories: Ancient Greece provides sharper questions than that. Firstly, how did Greeks tell their different types of stories? Oral and written means certainly, but also through visual and decorative arts such as drama, pottery, sculpture and inscription. Secondly, how do we find out about the Ancient Greeks? Reading whole stories from different sorts of texts is one way, but constructing them from fragmented evidence is another. Archaeologists use science and materials to tell stories and ask questions, often from minimal evidence. Seven- and 8-year-old children, meanwhile, are fascinated with the idea of what is underground, hidden or mysterious and can be enticed into archaeology by the attraction of its puzzles. Thirdly, how do different sorts of stories compare? The science of archaeology and the literature of Ancient Greece met in Schliemann's nineteenth century search for the site of Troy. Similarly the visual evidence on decorated pots contains both historical material and literary tellings; evidence of 'actual' Ancient Greek clothes, musical instruments and buildings but also of the significance of myths and stories. Children study Ancient Greece not just to hear and read Ancient Greek stories in their own right, but also to play with a question which will persist throughout their education: how many different types of story can human ingenuity tell?

Y4 term 1 requires children to experience 'historical stories and short novels . . . and a range of text-types from reports and articles in newspapers and magazines' (DfEE, 1998, p.38). Studying The Victorians gives chances for both to occur. Numerous children's authors have written about this period and can be read in total or extract. This stands in its own right as an experience but also prompts other questions. How were these historical stories researched? How accurate are they? Can we tell or write our own, especially after immersing ourselves in evidence from the period? Sometimes this will be in the form of newspaper reports and articles, a second stimulus. Most areas had a Victorian newspaper from which stories or extracts can be simplified or given to children as a whole text. These can be compared with modern local newspapers, used as a source to find out more about the Victorians or serve as a stimulus for imaginative storytelling. Often the smaller details of a newspaper suggest the most eye-catching stories: an advert, a letter, an obituary, the laconic report of a house burning or a child being convicted. These lead onto other texts: scrutinizing the reputations of real Victorian local heroes or heroines, reading from statues and plaques and plinths, savouring the poetry and singing the hymns from which Victorian children learnt.

Y5 term 2 requires children to experience 'traditional stories, myths, legends, fables from a range of cultures, longer classic poetry including narrative poetry . . . non-chronological reports . . . explanations . . . use content from other subjects' (DfEE, 1998, p.46). Studying the poetry, literature and history of the Romans, Anglo-Saxons and Vikings provides multiple opportunities for informative and imaginative-led learning, and, despite general imbalanced sources, need not ignore wider issues such as gender or class. For instance, telling the story of these societies only through their fictional stories and poems might lead to rather strange images of life for everyday men as well as women and children. Equally, to ignore a society's myths, heroic poetry or folk tales in favour only of supposedly factual material would be unhistorical as well as dull. Children enjoy both types of source and can start to compare different messages to ask interesting questions. What were Vikings really like? Who were and are the English? How can we find out about Anglo-Saxon or Roman women and children?

Y6 term 1 requires children to experience 'classic fiction, poetry and drama including where appropriate study of a Shakespeare play . . . adaptations of classics on film/TV . . . autobiography and biography, diaries, journals, letters, anecdotes, records of observations, etc., which recount experiences and events' (DfEE, 1998, p.50). Studying the life stories of Shakespeare and Elizabeth I, or comparing the imaginary witches of Macbeth and the 'real' witches of Elizabethan England, or reading individual stories about a range of Tudor people gives children experience of a wide range of difficult texts and historical questions. One of the most important concerns interpretations, a thread persistent in all storied materials but with which this example can lead. How have such stories differently represented the past? Why do people tell different stories about the same event or person? How do different interpretations of the same events arise and how can they be judged?

At the time of writing materials and policies are being piloted to extend the National Literacy Strategy into the secondary age range: yet curriculum opportunities have always existed for utilizing story, promoting literacy and developing historical thinking with older students. In the author's experience rigid timetabling and strong cultures of departmentalism, especially in larger secondary schools, have tended to maintain curriculum boundaries rather than encourage interdisciplinary planning. Stories have therefore often been identified with the subject of English, the teaching methods of drama or the idiosyncrasies of particularly talented teachers, rather than seen as a general pedagogic device. It need not be so, especially if, rather than *transmitting* NC requirements, teachers can confidently *translate* them.

This can be achieved by planning through key 'questions, stories and texts for reading and writing'. Such an approach has helped the author to synthesize NC content and objectives into stories that are meaningful and enquiry-led. For instance, the following examples translate broad NC history (DFE, 1995a) and English (DFE, 1995b) guidelines for Years 7, 8 and 9. When synthesized with practical teaching strategies outlined in previous chapters and elsewhere (e.g. Counsell, 1997) they form the basis for planning a curriculum which, although it may be storied, still develops disciplines in structured and principled ways:

Y7 Medieval People (*Medieval Realms*, NC history) NC English reading factual and informative texts (DFE, 1995b, p.21). Writing narrative and discursive, argumentative, persuasive and other non-fictional writing (ibid, pp.23–4).

Key Questions	*Key Stories*	*Key Reading Texts*	*Key Writing Texts*
Why did the Normans invade England?	The Battle of Hastings; The Domesday Book.	Film and video; textbooks; Simplified extracts (entries for local and distant places).	Storyboard; comparing films; stories by villagers and commissioners.
What sort of places were castles then — and now?	A local example or the Tower of London.	Guide books; information books; encyclopaedias.	Modern advertising text for tourists; descriptive text at time of building.
How did Medieval people think?	The evolution of the English language. The Great Pestilence.	Original examples (e.g. Anglo-Saxon, Latin, Chaucer). Chronicles; novels.	Research report on a medieval writer (e.g. scribe, nun, lady); medieval and modern explanations of the disease.
What did medieval people believe?	A local church or cathedral.	Buildings; sites and information texts; textbooks.	Medieval visions of heaven and hell.
How did ordinary medieval people live?	A real or imagined family.	Textbooks; novels; information books; videos.	Description of a house for a story; family tree; diaries.

Y8 Stories of Power, *Stories of Poverty* (*Britain 1750–1900*, NC history) NC English reading plays, novels, short stories and poems, some pre-1900 (DFE, 1995b, pp.19–21), introduced to non-fiction and media texts, synthesize information. In writing develop narrative use from knowledge of 'story structure . . . description . . . organization . . . characters and relationships' (ibid, p.23).

Key Questions	Key Stories	Key Reading Texts	Key Writing Texts
How did factories and railways start?	The Industrial Revolution.	Paintings; buildings; photographs; poems; textbooks.	A romantic description; recount or explanation.
What travelled between Britain and its colonies?	Emigration; immigration; trade (e.g. slavery); culture (e.g. ideas, goods).	Posters; letters; pictures; information books.	Poetry (departure, arrival); persuasive writing (slavery).
Which year tells the most important story?	1776 (America); 1789 (France); 1815 (Europe).	Short stories; textbooks; information books; novels.	Discussion; short story.
Did crowds ever change things?	Popular protest in England and elsewhere.	Songs; pamphlets; newspaper accounts.	Report (eyewitness); short story.
What was it like to be poor?	Dispossessed or marginalized people.	Novels; poems; reports.	Synthesized; discursive account; poem or song.
What was it like to be powerful?	Influential figures (e.g. intellectuals; industrialists; leaders).	Letters; biographies.	Biographic recount; letter/report.

Y9 Heaven or Hell? The Story of the Twentieth Century (*The 20th Century World*, NC history). In NC English read and write in many forms e.g. 'notes, diaries, personal letters, formal letters, chronological accounts, reports, pamphlets, reviews, essays, advertisements, newspaper articles, biography, autobiography, poems, stories, playscripts, screenplays' (DFE, 1995b, p.23).

Key Questions	Key Stories	Key Reading Texts	Key Writing Texts
How did they fight World War I?	Life in the trenches; the home front.	Letters; poems; films; memorials.	Personal letters; persuasive account.
Why did they fight World War I?	The assassination; the arms race; nations as ideas.	Textbooks; newspaper stories; propaganda; songs.	Explanatory essay.
Would I have followed Hitler?	Hitler's biography; the growth of Nazism.	Biographies; textbooks; information books.	Diaries; recount.
What did it mean to be Jewish?	Jewish beliefs and culture; The Holocaust.	Diaries and accounts; propaganda materials; autobiographies; films; dramas.	A personal response.

Key Questions	Key Stories	Key Reading Texts	Key Writing Texts
What did it mean to be Japanese?	The Japanese empire; the atomic bombs.	Accounts; textbooks; letters; articles.	Newspaper reports; discursive analysis.
What has changed since 1945?	History through the eyes of individuals.	Interviews; biography; autobiography; environments.	Biography of a person interviewed; biography of a leading figure.

So, reader, how do we finish? Perhaps, as Zipes suggested, the role model for the storyteller should be the little child in Andersen's *The Emperor's New Clothes*: 'The child's single sentence is a provocative and subversive story, for it liberates the bystanders to think and speak what they have been repressing' (1995, p.224). Which brings me to a final question of you: does your curriculum have any clothes?

And that, of course, is something you might ask your children . . .

In Practice (9)

Oral Storytelling and the National Literacy Strategy

The following examples outline how oral storytelling can enhance teaching within the National Literacy Strategy and Framework. It is almost entirely the work of Rachael Sutherland (Literacy Consultant, Greenwich LEA) a teacher and adviser who is participating in the teaching-as-storytelling project in Greenwich.

Does the starting point for shared reading and writing always have to be printed text, or can teachers and children draw upon oral versions of stories? From reception onwards, the learning objectives of the National Literacy Framework (DfEE, 1998) emphasize the comparison of oral and written forms. This is only possible when children are experiencing oral tellings of stories as well as written texts. Similarly, children are expected to retell stories in a variety of ways. This will often be easiest through experimenting with oral retellings, before moving on to writing. Through this children gain confidence in controlling the structure of a story and are able to experiment with different forms of language and a wider vocabulary. The literacy hour also requires teachers to model skills and strategies for the class, which groups and individuals then try for themselves. Teaching can begin by presenting oral versions, to which children respond. These can be compared with written texts and class versions of stories produced in shared writing. Children at this point can incorporate features of what they have read and heard in their own writing, supported by teacher feedback to increase their independent and confident use of a particular form.

Typical Oral and Storytelling Activities Within the Literacy Hour:

- teacher telling stories e.g. to introduce a particular form, to provide material for comparison;
- teacher initiating discussion of main features, e.g. plot structure, main events, characters, settings;
- teacher incorporating range of traditional story openings/ endings into tellings;
- teacher demonstrating planning to support retelling, e.g. storyboards, maps, plans, notes;
- teacher telling different version of traditional stories, stories from different cultures, etc.;
- teacher modelling the comparison of oral and written forms, drawing attention to similarities and differences;
- teacher modelling performance, leading discussion on, e.g. use of rich and varied vocabulary;
- teacher leading discussion on meanings beyond the literal, e.g. motivation for characters' actions, details that have been included or excluded from the story;
- children retelling stories they have heard . . . to individuals, groups, class;
- children telling stories they have devised . . . to individuals, groups, class;
- children devising and telling stories based on 'real life' experiences (own, historical, contemporary);
- children exploring story structure;
- children planning story structure, e.g. props, storyboards, story sticks, maps, notes, etc.;
- children preparing and performing retellings using story language and wide vocabulary; giving feedback to one another on aspects of performance;
- children using features of stories they have heard in retelling and written versions, e.g. characters, aspects of plot, repetitions, story openings and endings, etc.;
- children visualizing aspects of stories they have heard, extending their use of descriptive language;
- children responding to stories, comparing, expressing and explaining preferences.

Drawing on the broader principles outlined above, the following grid identifies specific opportunities for using storytelling within the NLS Framework's Fictional and Poetic range of texts (DfEE, 1998). Learning objectives are described, all of which can be taught through told stories. These can draw from a range of content (e.g. English, history, RE, geography, science, multicultural education) and could be extended to include non-fiction stories — though space precludes us from describing those possibilities here.

Range — Fiction and poetry elements to introduce and explore through storytelling	Text Level Objectives — objectives to model and reinforce through storytelling strategies
Reception Traditional stories, rhymes; stories and rhymes with predictable and patterned language	4 to notice the difference between spoken and written forms through retelling known stories; to compare 'told' versions with what the book 'says'; 5 to understand how storybook language works and to use some formal elements when retelling stories; 7 to use knowledge of familiar texts to re-enact or retell to others, recounting the main points in correct sequence; 9 to be aware of story structures, e.g. actions/reactions, consequences, and the ways that stories are built up and concluded.
Year 1 Term 1 Stories and rhymes with predictable and patterned language	**Term 1** 3 to notice the difference between spoken and written forms retelling known stories; compare oral versions with the written text; 5 to describe story settings and incidents and relate them to their own experience and that of others;

6 to recite stories and rhymes with predictable and repeating patterns, extemporizing on patterns orally by substituting words and phrases, extending patterns, inventing patterns and playing with rhyme;

7 to re-enact stories in a variety of ways, e.g. through role play, using dolls or puppets;

10 to use rhymes and patterned stories as models for their own writing.

Year 1 Term 2
Traditional stories, fairy stories, stories and poems with predictable patterned language from range of cultures (including playground chants, etc.)

Term 2

4 to retell stories, giving the main points in sequence and to notice differences between the written and spoken forms in retelling, e.g. by comparing oral versions with the written text;

5 to identify and record some key features of story language from a range of stories, and to practise reading and using them, e.g. in oral retellings;

6 to identify and discuss a range of story themes, and to collect and compare;

7 to discuss reasons for, or causes of, incidents in stories;

8 to identify and discuss characters;

10 to identify and compare basic story elements, e.g. beginnings and endings in different stories;

14 to represent outlines of story plots using e.g. captions, pictures, arrows to record main incidents in order;

16 to use some elements of known stories to structure own writing.

Year 1 Term 3
Stories about fantasy worlds

Term 3

3 to notice the difference between spoken and written forms through retelling known stories; compare oral versions with the written text;

	5 to retell stories, to give the main points in sequence and to pick out significant incidents; **6** to prepare and retell stories orally, identifying and using some of the more formal features of story language; **10** to compare and contrast preferences and common themes in stories and poems.
Year 2 Term 1 Stories . . . with a variety of familiar settings	**3** to be aware of the difference between spoken and written language through comparing oral recounts with text; make use of formal story elements in retelling; **4** to understand time and sequential relationships in stories; **5** to identify and discuss reasons for events in stories, linked to plot; **6** familiarize story themes and link to own experiences; **10** to use story structure to write about own experience in same/ similar form.
Year 2 Term 2 Traditional stories; stories and poems from other cultures; stories and poems with predictable and patterned language	**Term 2** **3** to discuss and compare story themes; **4** to predict story endings/ incidents; **5** to discuss story settings; **7** to prepare and retell stories individually and through role-play in groups, using narrative and dialogue from text; **13** to use story settings from reading (when writing); **14** to write character profiles using key words and phrases that describe or are spoken by characters in the text.

Year 2 Term 3 Texts with language play, e.g. riddles, tongue-twisters, humorous verse and stories	**Term 3** **3** to notice the difference between spoken and written forms through retelling known stories; compare oral versions with the written text; **6** to read, respond imaginatively, recommend and collect examples of humorous stories, extracts, poems; **8** to discuss meanings of words and phrases that create humour.
Year 3 Term 2 Myths, legends, fables, parable; traditional stories; oral and performance poetry from different cultures	**Term 2** **1** to investigate the styles and voices of traditional story language — collect examples, e.g. story openings and endings; **2** to identify typical story themes, e.g. trials and forfeits, good over evil, weak over strong, wise over foolish; **3** to identify and discuss main and recurring characters, evaluate their characters and justify their views; **4** to choose and prepare poems for performance, identifying appropriate expression, tone, volume and use of voices and other sounds; **7** to describe and sequence key incidents in a variety of ways, e.g. by listing, charting, making simple storyboards; **9** to write a story plan for own myth, fable or traditional tale, using story theme from reading but substituting different characters or changing the setting;

	10 to write alternative sequels to traditional stories using same characters and settings, identifying typical phrases and expressions from story and using these to help structure the writing; **11** to write new or extended verses for performance based on models of 'performance' and oral poetry read, e.g. rhythms, repetition.
Year 3 Term 3 adventure and mystery stories; poetry that plays with language, word puzzles, puns, riddles	**Term 3** **1** retell main points of story in sequence; **3** to distinguish between first and third person accounts; **4** to consider credibility of events, e.g. by selecting some real life adventures either written or retold as stories and comparing them with fiction; **7** to select, prepare, read aloud and recite by heart poetry that plays with language or entertains; **12** to write a first person account, e.g. write a character's own account of incident in story read.
Year 4 Term 1 Historical stories and short novels	**Term 1** **3** to explore chronology in narrative using written or other media texts, by mapping how much time passes in the course of the story; **4** to explore narrative order; identify and map out the main stages of the story; **10** to plan a story identifying the stages of its telling; **12** to write independently, linking own experience to situations in historical stories, e.g. how would I have responded? what would I do next?

Year 4 Term 3 Stories from other cultures	**Term 3** 2 to read (hear) stories from other cultures, by focusing on, e.g. differences in place, time, customs, relationships; to identify and discuss recurring themes where appropriate.
Year 5 Term 2 Traditional stories, myths, legends, fables from a range of cultures; longer classic poetry including narrative poetry	**Term 2** 1 to identify and classify the features of myths, legends and fables; 2 to investigate different versions of the same story in print or on film (or told versions), identifying similarities and differences, recognize how stories change over time, differences of culture and place that are expressed in stories; 3 to explore similarities and differences between oral and written story telling; 11 to write own versions of legends, myths and fables, using structures and themes identified in reading (telling).
Year 5 Term 3 Stories from a variety of cultures and traditions; choral and performance poetry	**Term 3** 1 to investigate a range of texts from different cultures, considering patterns of relationships, social customs, attitudes and beliefs; consider and evaluate these features in relation to their own experience; 2 to identify the point of view from which the story is told and how this affects the reader's response; 3 to change the point of view, e.g. tell incident or describe a situation from the point of view of another character or perspective; 4 to read, rehearse and modify performance poetry.

Year 6 Term 1

Classic fiction, poetry and drama by long-established authors including Shakespeare; adaptations of classics on film/TV

Term 1

1 to compare and evaluate . . . print and the TV/film version;
2 to take account of viewpoint in a novel through identifying the narrator, explaining how this influences the reader's view of events, explaining how events might look from a different point of view;
3 to articulate personal responses to literature, identifying how and why a text affects the reader;
5 to contribute constructively to shared discussion about literature;
6 to manipulate narrative perspective by writing in the voice and style of a text, producing a modern retelling, writing a story with two different narrators;
7 to plan quickly and effectively the plot, characters and structure of their own narrative writing.

Year 6 Term 2

Longer established stories and novels from more than one genre . . . Range of poetic forms, e.g. kennings, limericks, riddles, cinquain, tanka, poems written in other forms (e.g. as adverts, letters, diary entries, conversations), free verse, nonsense verse

Term 2

1 to understand aspects of narrative structure, e.g. how chapters or paragraphs are linked together . . . how authors handle time . . . how the passing of time is conveyed;
3 to recognize how poets manipulate words for their quality of sound, connotations and multiple layers of meaning;
5 to analyse how messages, moods, feelings and attitudes are conveyed in poetry;
7 to identify key features of different types of literary text;
11 to write own story using e.g. flashbacks, or story within a story.

References

AAM (1950) *The Teaching of History*, Issued by the Incorporated Association of Assistant Masters in Secondary Schools, Cambridge: CUP.

ABBS, P. (ed.) (1993) 'Socratic education — aspects of education', *Journal of the Institute of Education*, **49**, University of Hull.

ABBS, P. (1994) *The Educational Imperative: A Defence of Socratic and Aesthetic Learning*, London: Falmer Press.

ALDRICH, R. (1991) *History in the National Curriculum*, London: Kogan Page.

ALDRICH, R. and DEAN, D. (1991) 'The historical dimension', in ALDRICH, R. (ed.) *History in the National Curriculum*, London: Kogan Page.

ALEXANDER, P., SCHALLERT, D. and HARE, V. (1991) 'Coming to terms: How researchers in learning and literacy talk about knowledge', *Review of Educational Research*, **61**, 3, pp.315–43.

ALEXANDER, R. (1995) *Versions of Primary Education*, London: Routledge.

ALLASON-JONES, L. (1989) *Women in Roman Britain*, London: British Museum Publications.

AMY, S. (1986) *Storytelling Rights — The Uses of Oral and Written Texts by Urban Adolescents*, Cambridge: CUP.

ANDERSON, G., HERR, K. and SIGRID NIHLEN, A. (1994) *Studying Your Own School: An Educator's Guide to Qualitative Practitioner Research*, California: Corwin Press.

ANDREETTI, K. (1993) *Teaching History from Primary Evidence*, London: David Fulton.

APPLE, M. (1993) *Official Knowledge: Democratic Education in a Conservative Age*, London: Routledge.

APPLEBEE, A. (1978) *The Child's Concept of Story*, Chicago, IL and London: Chicago University Press.

APPLEBY, J., HUNT, L. and JACOB, M. (1994) *Telling The Truth about History*, New York: W.W.Norton & Co.

APPLEYARD, M. (1992) *The Times* 29 July 1992 'Losing faith in learning', p.12.

ARIS, M. and BROOKS, R. (1993) *The Effective Teaching of History*, London: Longman.

AUBREY, C. (1995) *The Role of Subject Knowledge in the Early Years of Schooling*, London: Falmer Press.

AUSTEN, J. (1818) *Northanger Abbey*, London: Penguin, 1972 ed.

BAGE, G. (1993) 'History at KS1 and KS2', *Curriculum Journal*, **4**, 2, Summer 1993, pp.269–82.

BAGE, G. (1995) 'Chaining the beast? An autobiographical examination by an advisory teacher of whether spoken storytelling and prompting can make school history's analytic transmission more educationally principled and powerful', UEA: Unpublished PhD.

BAGE, G. (1996) 'Viewpoint', *Primary History*, **13** (June), p.4.

BAGE, G., GROSVENOR, J. and WILLIAMS, M. (1999) 'Curriculum planning: Prediction or response?', *Curriculum Journal*, **10**, pp.49–69.

BAKER, K. (ed.) (1988) *English History in Verse*, London: Faber and Faber.

BALDWIN, G. (1994) 'A Dearing opportunity: History teaching and moral education', *Teaching History*, **76**, pp.29–32.

BALL, S. (1994) *Education Reform: A Critical and Post-structuralist Perspective*, Buckingham: OUP.

BARKER, P. (1995) *The Ghost Road*, London: Penguin.

BARNES, D. (1976) *From Communication to Curriculum*, London: Penguin.

BARNES, D. (1988) 'The politics of oracy', in MACLURE, M. (ed.) *Oracy Matters*, Milton Keynes: OUP.

BASSEY, M. (1995) *Creating Education Through Research*, Newark NJ: Kirklington Moor Press.

BEARE, H. and SLAUGHTER, R. (1993) *Education for the Twenty-First Century*, London: Routledge.

BEN-PERETZ, M. (1995) *Memory and the Teacher's Account of Teaching*, Albany, NY: SUNY Press.

BENNETT, N. (1976) *Teaching Styles and Pupil Progress*, London: Open Books.

BERRILL, D. (1988) 'Anecdote and the development of oral argument in 16-year-olds', in MACLURE, M. (ed.) *Oracy Matters*, Milton Keynes: OUP.

BLUNT, C. (1969) 'The Saint Edmund memorial coinage', in the *Proceedings of the Suffolk Institute of Archaeology*, **31**, 3.

BLYTH, J. (1985) *Place and Time with Children Five to Nine*, London: Routledge, 1988 edition.

BLYTH, J. (1988) *History 5–9*, London: Hodder and Stoughton.

BLYTH, J. (1989) *History in Primary Schools*, Milton Keynes: OUP.

BOLTON, G. (1994) 'Stories at work. Fictional-critical writing as a means of professional development', *British Educational Research Journal*, **20**, 1, pp.55–68.

BOWKER (1993) THES 8 January 1993, p.19.

BRETT, S. (1989) *The Faber Book of Diaries*, London: Faber and Faber.

BRIGHOUSE, T. (1994) *Times Educational Supplement*, 27 April 1994, Review Section 2, p.1.

BROOKFIELD, S. (1995) *Becoming a Critically Reflective Teacher*, San Francisco, CA: Jossey-Bass.

BROOKS, G., GORMAN, T., HARMAN, J., HUTCHISON, D. and WILKIN, A. (1996) *Family Literacy Works*, London: The Basic Skills Agency.

BROOKS, R. (1991) *Contemporary Debates in Education — An Historical Perspective*, London: Longman.

BRUNER, J. (1960) *The Process of Education*, Cambridge, MA: Harvard University Press, 1965 edn.

BRUNER, J. (1986) *Actual Minds, Possible Worlds*, London: Harvard University Press.

BRUNER, J. (1990) *Acts of Meaning*, Cambridge, MA: Harvard University Press.

BUCKLEY, J. (1994) 'A critique of Kieran Egan's theory of educational development', *Journal of Curriculum Studies*, **26**, 1, pp.31–43.

BURSTALL, S. (1895) 'History teaching in schools', *The Journal of Education*, June, p.381.

BUTT, R., TOWNSEND, D. and RAYMOND, D. (1990) 'Bringing reform to life: Teacher's stories and professional development', *Cambridge Journal of Education*, **20**, 3, pp.255–68.

CALDER, A. and SHERIDAN, D. (1985) *Speak for Yourself*, Oxford: Oxford University Press.

CALLCOTT (LADY) (1878) *Little Arthur's England*, London: John Murray.

CANNADINE, D. (1989) *The Pleasures of the Past*, London: Collins.

CARLYLE, T. (1841) *On Heroes, Hero-Worship and the Heroic in History*, Goldberg, M. (ed.) (1993) Berkeley CA: University of California Press.

CARR, W. and KEMMIS, S. (1986) *Becoming Critical*, Lewes: Falmer Press.

CASEY, K. (1994) *Sunday Times*, Culture 13 March 1994, p.9.

CASEY, K. (1990) 'Teacher as mother: Curriculum theorizing in the life histories of contemporary women teachers', *Cambridge Journal of Education*, **20**, 3, pp.301–20.

CHANCELLOR, V. (1970) *History for their Masters: Opinion in the English History Textbook, 1800–1914*, Bath: Adams and Dart.

CHANG, J. (1991) *Wild Swans*, London: Harper Collins, paperback edition of 1993.

CHARLES, R. (1895) 'History teaching in schools', *Journal of Education*, **311**, June, p.379.

CLAIRE, H. (1996) *Reclaiming Our Pasts: Equality and Diversity in the Primary History Curriculum*, Stoke: Trentham Books.

CLANDININ, D. and CONNELLY, F. (1987) 'Teachers' personal knowledge: What counts as "personal" in studies of the personal', *Journal of Curriculum Studies*, **19**, 6, pp.487–500.

CLANDININ, D. and CONNELLY, F. (1990) 'Narrative, experience and the study of curriculum', *Cambridge Journal of Education*, **20**, 3, pp.241–53.

CLANDININ, D. and CONNELLY, F. (1995) *Teachers' Professional Knowledge Landscapes*, New York: Teachers' College Press.

CLEAVER, L. (1985) 'Oral history at Thurston Upper School', in *Oral History Journal*, **13**, 1.

COLLICOTT, S. (1986) *Connections: Haringey Local–National–World Links*, London: Haringey Multi-Cultural Curriculum Support Group.

COLLICOTT, S. (1992) 'Who is forgotten in HSU Britain since the 1930s?', *Primary Teaching Studies*, **6**, 3, pp.252–68.

COLLICOTT, S. (1993) 'A way of looking at history: Local–national–world links', *Teaching History*, July, pp.18–23.

COLLINGWOOD, R. (1939) *An Autobiography*, Oxford: OUP, 1978 edn.

COLLINGWOOD, R. (1946) *The Idea of History*, Oxford: OUP, 1961 edn.

CONNELLY, F. and CLANDININ, D. (1986) 'On narrative method, personal philosophy, and narrative unities in the story of teaching', *Journal of Research in Science Teaching*, **23**, 4, pp.293–310.

CONNELLY, F. and CLANDININ, D. (1990) 'Stories of experience and narrative enquiry', *Educational Researcher*, **19**, 5, pp.2–14.

COOK, E. (1976) *The Ordinary and the Fabulous*, London.

COOLING, M. (1994) *Faith in History — Ideas for RE, History and Assembly in the Primary School*, Guildford: Eagle Publishing.

COOPER, H. (1991) *Times Educational Supplement*, 6 December 1991, 'Alice and Wonderland', p.48.

COOPER, H. (1992) *The Teaching of History*, London: Fulton Press.

COOPER, H. (1995a) *History in the Early Years*, London: Routledge.

COOPER, H. (1995b) *The Teaching of History in Primary Schools*, London: David Fulton, 2nd edn.

CORBETT, P. (1990) 'An historical view of the relationship between reading and writing', in ENOS, R. (ed.) *Oral and Written Communication — Historical Approaches*, London: Sage Publications.

CORTAZZI, M. (1991) *Primary Teaching How It Is: A Narrative Account*, London: David Fulton Publishers.

CORTAZZI, M. (1993) *Narrative Analysis*, London: Falmer Press.

COUNSELL, C. (1997) *Analytical and Discursive Writing at Key Stage 3*, London: Historical Association.

COX, K. and HUGHES, P. (1990) *Early Years History: An Approach Through Story*, Liverpool: Liverpool Institute of Higher Education.

CRAMER, I. (1993) 'Oral history: Working with young children', *Teaching History*, **71**, April.

CRAWFORD, K. (1995) 'A history of the right: The battle for control of national curriculum history 1989–1994', *British Journal of Educational Studies*, **43**, 4, pp.433–56.

CRAWFORD, K. (1998) 'The teaching and learning of primary history', *Teaching History*, **90**, pp.33–7.

CRITES, S. (1986) 'Story time: Recollecting the past and projecting the future', in SARBIN, T. (ed.), *Narrative Psychology*, New York: Praeger.

CROLL, P. (1996) *Teachers, Pupils and Primary Schooling*, London: Cassell.

CROSSLEY-HOLLAND, K. (1982) *The Anglo-Saxon World*, Woodbridge: Boydell Press.

CULLINGFORD, C. (1995) *The Effective Teacher*, London: Cassell.

CULPIN, C. (1984) 'Language, learning and thinking skills in history', *Teaching History*, June, pp.24–9.

CUNNINGHAM, P. (1992) 'Teachers' professional image and the press 1950–1990', in *History of Education*, **21**, 1, pp.37–56.

CURTIS, S. (1994) 'Communication in history', in *Teaching History*, **77**, pp.25–30.

CURTIS, S., BARDWELL, S. (1994) 'Access to history', in BOURDILLON, H. (ed.), *Teaching History*, London: Routledge.

DADDS, M. (1993) 'The changing face of topic work in the primary curriculum', *Curriculum Journal*, **4**, 2, pp.253–68.

DADDS, M. (1995) *Passionate Enquiry and School Development — A Story about Teacher Action Research*, London: Falmer Press.

DE MARCO, N. (1992) *Times Educational Supplement*, 11 December 1992, History Special Report, 'Dogma alert', p.1.

DEARING, R. (1993) *The National Curriculum and Its Assessment*, Interim Report, NCC/ SEAC, York and London: NCC.

DENSCOMBE, M. (1984) 'Interviews, accounts and ethnographic research on teachers', in HAMMERSLEY, M. (1984) *The Ethnography of Schooling*, Humberside: Nafferton Books.

DERRY, T.K. (1987) 'The martyrdom of St Edmund AD 869', *Norsk Historisk Tidsskrift*, **66**, pp.157–63.

DES (1989) *Aspects of Primary Education: The Teaching and Learning of History and Geography*, London: HMSO.

DES (1990a April) *National Curriculum History Working Group*, London: HMSO.

DES (1990b July) *History for Ages 5 to 16. Proposals of the Secretary of State*, London: HMSO.

DES (March 1991) *History in the National Curriculum — Final Order*, London: HMSO.

DEUCHAR, S. (1993) 'Curious incident of the dog in history', *Times Educational Supplement*, 12 March, p.5.

DFE (1995a) *History in the National Curriculum*, London: HMSO.

DFE (1995b) *English in the National Curriculum*, London: HMSO.

DfEE (1997a) *The Implementation of the National Literacy Strategy*, London: DfEE.

DfEE (1997b) *Excellence in Schools*, London: DfEE.

DfEE (1998) *The National Literacy Strategy — Framework for Teaching*, London: DfEE.

DICKINSON, A. (1991) 'Assessing, recording and reporting children's achievements', in ALDRICH, R. (ed.) *History in the National Curriculum*, London: Kogan Page.

DIXON, J. (1988) 'Oral exchange: A historical review of the developing frame', in MACLURE, M. (ed.) *Oracy Matters*, Milton Keynes: OUP.

DODGSON, E. (1984) 'From oral history to drama', *Oral History Journal*, **12**, 2, pp.47–53.

DONALDSON, M. (1978) *Children's Minds*, London: Fontana.

DRUMMOND, M.J. and MCCLAUGHLIN, C. (1994) 'Teaching and learning — The fourth dimension', in BRADLEY, H., CONNER, C. and SOUTHWORTH, G. (eds) *Developing Teachers Developing Schools*, London: David Fulton.

DUNAWAY, D. and BAUM, W. (eds) (1996) *Oral History: An Inter-disciplinary Anthology*, 2nd edition, Walnut Creek: Altamira Press.

DUPASQUIER, P. (1988) *Wagons West*, London: Penguin.

ECO, U. (1980) *The Name of the Rose*, Transl. WEAVER, W. (1984), London: Picador.

EDWARDS, A. and WESTGATE, D. (1987) *Investigating Classroom Talk*, Lewes: Falmer Press.

EDWARDS, A. (1978) 'The language of history and the communication of historical knowledge', in DICKINSON, A. and LEE, P., *History Teaching and Historical Understanding*, London: Heinemann.

EGAN, K. (1979) *Educational Development*, New York: Oxford University Press.

EGAN, K. (1983) *Education and Psychology*, New York: London, Methuen.

EGAN, K. (1985) 'Teaching as story-telling: A non-mechanistic approach to planning teaching', *Journal of Curriculum Studies*, **17**, 4, pp.397–406.

EGAN, K. (1989) *Teaching as Storytelling*, New York: London, Routledge 1989 ed.

EGAN, K. (1990) *Romantic Understanding*, London: Routledge, 1992 ed.

ELBAZ, F. (1991) 'Research on teacher's knowledge', *Journal of Curriculum Studies*, **23**, 1, pp.1–19.

ELINOR, G. (1992) 'Stolen or given?', in *Oral History*, **20**, 1, pp.78–9.

ELLIOTT, J. (1991) *Action Research for Educational Change*, Milton Keynes: OUP.

ELTON, G. (1967) *The Practice of History*, Sydney: London, Methuen.

ENGLISH HERITAGE EDUCATION SERVICE (1996) *Talkin' Roman* — A video for KS2/3 Halesworth: Suffolk Films.

FAIRCLOUGH, J. (1994) *A Teacher's Guide to History Through Role Play*, London, English Heritage.

FARMER, A. (1990) 'Story-telling in History', *Teaching History*, January, pp.17–23.

FINEMAN, J. (1989) 'The history of the anecdote: Fiction and fiction', in ARAM, V. (ed.) *The New Historicism*, ed. H. ARAM VEESER, London: Routledge.

FINES, J. (1975) 'The narrative approach', *Teaching History*, **4**, 14, pp.97–104.

FINES, J. (1987) 'Making sense out of the content of the history curriculum', in PORTAL, C. (ed.) *The History Curriculum for Teachers*, Lewes: Falmer Press.

FINES, J. (1997) 'Truth and the imagination: A little investigation in three fits', in DAVIS, D. (ed.) *Interactive Research in Drama in Education*, Stoke: Trentham Books.

FINES, J. and NICHOL, J. (1997) *Teaching Primary History*, London: Heinemann.

FINES, J. and VERRIER, R. (1974) *The Drama of History*, London: New University Education.

FISCHER, D. (1970) *Historians Fallacies — Towards a Logic of Historical Thought*, London: Routledge and Kegan Paul.

FOUCAULT, M. (1977) *Language, Counter-Memory, Practice*, BOUCHARD, D. (ed.) Oxford: Blackwell.

FRATER, G. (1988) 'Oracy in England — a new tide', in MACLURE, M. (ed.) *Oracy Matters*, Milton Keynes: OUP.

FULLAN, M. (1991) *The New Meaning of Educational Change*, London: Cassell.

GABELLA, M. (1994) 'The art(s) of historical sense', *Journal of Curriculum Studies*, **27**, 2, pp.139–63.

GALTON, M. (1995) *Crisis in the Primary Classroom*, London: Fulton.

GARDINER, J. (ed.) (1990) *The History Debate*, London: Collins and Brown.

GARDNER, P. (1993) 'Uncertainty, teaching and personal autonomy', *Cambridge Journal of Education*, **23**, 2, pp.155–71.

GERRARD, R. (1990) *Mik's Mammoth*, London: Victor Gollancz.

GERRARD, R. (1996) *Wagons West*, London: Victor Gollancz.

GILLBORN, D. (1997) 'Racism and reform: New ethnicities/old inequalities?', *British Educational Research Journal*, **23**, 3, pp.345–60.

GOALEN, P. and HENDY, L. (1994) 'History through drama', *Curriculum*, **15**, 3, pp.147–62.

GOLD, K. (1994) 'Perspective: We are family', *Times Higher Education Supplement*, 1 July, pp.17–18.

GOODSON, I. (1978) 'New views of history: From innovation to implementation', in DICKINSON, A., GARD, A. and LEE, P.J. (eds) *History Teaching and Historical Understanding*, London: Heinemann.

GOODSON, I. (1984) 'The use of life histories in the study of teaching', in HAMMERSLEY, M. (ed.) *The Ethnography of Schooling*, Humberside: Nafferton Books.

GOODSON, I. (1995) 'The story so far', *International Journal of Qualitative Studies*, **8**, 1, pp.89–98.

GRAHAM, R. (1991) *Reading and Writing the Self: Autobiography in Education and the Curriculum*, New York: Teachers College Press.

GRAINGER, T. (1997) *Traditional Storytelling in the Primary Classroom*, Leamington Spa: Scholastic.

GREENWAY, D. and SAYERS, J. (1989) transl. *Jocelin of Brakelond Chronicle of the Abbey of Bury St Edmunds*, Oxford: World Classics, OUP.

GREGORY, E. (1994) 'Cultural assumptions and early years' pedagogy: The effect of the home culture on minority children's interpretation of reading in school', *Language, Culture and Curriculum*, **7**, 2, pp.111–23.

GRUMET, M. (1990) 'Retrospective: Autobiography and the analysis of educational experience', *Cambridge Journal of Education*, **20**, 3, pp.321–5.

GUDMUNDSDOTTIR, S. (1990) 'Four case studies of social studies teaching', in DAY, C., POPE, M. and DENICOLO, P. (eds), *Insight into Teachers' Thinking and Practice*, London: Falmer Press, p.117.

GUDMUNDSDOTTIR, S. (1991) 'Story-maker, story-teller: Narrative structures in curriculum', *Journal of Curriculum Studies*, **23**, 3, pp.207–18.

GUDSMUNDOTTIR, S. (1995) 'The narrative nature of pedagogical content knowledge', in MCEWAN and H. EGAN, K. (eds) *Narrative in Teaching, Learning and Research*, New York: Teachers' College Press.

HAMILTON, D. (1989) *Towards a Theory of Schooling*, Lewes: Falmer Press.

HAMILTON, D. (1990) *Learning about Education*, Milton Keynes: OUP.

HAMILTON, W. (1971) transl. *Plato The Gorgias*, London: Penguin.

HAMMERSLEY, M. (1984) 'Reflexivity and naturalism in ethnography', in HAMMERSLEY, M. (ed.) *The Ethnography of Schooling*, Humberside: Nafferton Books.

HAMMERSLEY, M. (1993) 'On the teacher as researcher', in *Educational Research — Current Issues*, Milton Keynes: OUP.

HANDFORD, S. (1951) transl. *Caesar The Conquest of Gaul*, London: Penguin.

HARDY, B. (1977) in MEEK, M., WARLOW, A. and BARTON, G. (eds 1977) 'Narrative as a Primary Act of Mind', London: Bodley Head.

HARNETT, P. (1993) 'Identifying progression in children's understanding: The use of visual materials to assess primary school children's learning in history', *Cambridge Journal of Education*, **23**, 2, pp.137–54.

HARLEN, W. (1996) 'Primary teachers' understanding in science and its impact in the classroom', Paper presented at BERA, September 1996.

HAVELOCK, E. (1986) *The Muse Learns to Write*, London: Yale University Press.

HAZAREESINGH, S. (1994) *Speaking about The Past. Oral History for 5–7 Year Olds*, Stoke: Trentham Books.

HOWARD, P. (1992) 'My life is not my own', *The Times*, 29 July 1992.

HUMANITIES CURRICULUM PROJECT (HCP) (1970) *An Introduction*, London: Heinemann 1983 ed. UEA: Norwich.

HENDY, L. (1996) 'With the wind behind you', in STYLES, M. et al. (1996) *Voices Off: Texts, Contexts and Readers*, London: Cassell.

HENIGE, D. (1982) *Oral Historiography*, Harlow: Longman.

HERODOTUS (1996) *The Histories*, DE SELINCOURT, A. MARINCOLA, J. (ed.) London: Penguin.

HERRNSTEIN-SMITH, B. (1981) 'Afterthoughts on narrative III' in MITCHELL, W. (ed.) *On Narrative*, Chicago, University of Chicago Press.

HERVEY, F. (1907) *Corolla Sancti Eadmundi: The Garland of St Edmund King and Martyr*, London: Murray.

HEWITT, M. and HARRIS, A. (1992) *Talking Time — A Guide to Oral History for Schools*, London: Tower Hamlets Education Committee.

HEXTER, J. (1972) *The History Primer*, London: Allen Lane.

HILTON, M. (1989) 'Stories and remembering', in STYLES, M. (ed.) *Collaboration and Writing*, Milton Keynes: OUP.

HILTON, M. (1996) *Potent Fictions: Children's Literacy and the Challenge of Popular Culture*, London: Routledge.

HISLAM, J. and JARVIS, T. (1992) 'Bridging the discipline gap: Linking story themes with science and technology', *Education 3–13*, June, pp.38–44.

HITCHCOCK, G. and HUGHES, D. (1989) *Research and the Teacher: A Qualitative Introduction to School-based Research*, London: Routledge.

HUGHES, E. (1977) *Times Educational Supplement*, 2 September, pp.11–13, 'Myth and education'.

HULL, R. (1986) *The Language Gap*, London: Methuen.

HUMPHRIES, S. (1981) *Hooligans or Rebels?*, Oxford: Blackwell.

HURCOMBE, L. (1997) 'A viable past in the pictorial present?', in MOORE, J. and SCOTT, E. (eds) (1997) *Invisible People and Processes — Writing Gender and Childhood into European Archaeology*, London: Leicester University Press.

HURT, J. (1972) *Aelfric*, New York: Twayne Publishers.

HUSBANDS, C. (1990) *Times Educational Supplement*, 6 July, letters page.

HUSBANDS, C. (1996) *What Is History Teaching? Language, Ideas and Meaning in Learning about the Past*, Buckingham: OUP.

IGNATIEFF, M. (1990) *Observer*, 28 October 1990.

ILEA (1979) '*Language and History*' in *Teaching History (1994)*, BOURDILLON, H. (ed.) London: Routledge.

ISENBERG, J. and JALONGO, M. (1995) *Teachers' Stories*, San Francisco, CA: Jossey-Bass.

JACKSON, D. (1983) 'Dignifying anecdote', in *English in Education*, **17**, 1, pp.7–21.

JACKSON, P. (1995) 'On the place of narrative in teaching', in McEWAN, H. and EGAN, K. (eds) *Narrative in Teaching, Learning and Research*, New York: Teachers' College Press.

JAMES, M. (1895) *On the Abbey of St Edmund at Bury*, Cambridge: Cambridge Antiquarian Society.

JENKINS, I. (1986) *Greek and Roman Life*, London: British Museum.

JENKINS, K. (1997) *The PostModern History Reader*, London: Routledge.

JENKINS, K. and BRICKLEY, P. (1990) 'Designer histories: A reply to Christopher Portal', *Teaching History*, July, pp.27–9.

JOHN, P. (1994a) 'The integration of research validated knowledge with practice: Lesson planning and the student history teacher', *Cambridge Journal of Education*, **24**, 1, pp.33–47.

JOHN, P. (1994b) 'Academic tasks in history classrooms', in *Research in Education*, **51**, pp.11–22.

JONES, P. (1988) *Lip Service: The Story of Talk in Schools,* Milton Keynes: OUP.

JOSIPOVICI, G. (1993) 'Life between inverted commas', *TES* 12 February 1993, p.9.

JOYCE, B., CALHOUN, E. and HOPKINS, D. (1997) *Models of Learning — Tools for Teaching*, Buckingham: OUP.

KEATING, M. (1910) *Studies in the Teaching of History*, London.

KING, C. (1988) 'The historical novel: An under-used resource', *Teaching History*, April 1988, pp.24–6.

KNIGHT, P. (1989) 'Empathy: Concept, confusion and consequences in a national curriculum', *Oxford Review of Education*, **15**, 1, pp.41–53.

KNIGHT, P. (1991) 'Teaching as exposure: The case of good practice in junior school history', *British Educational Research Journal*, **17**, 2, pp.129–40.

KNIGHT, P. (1993) *Primary Geography, Primary History*, London: Fulton.

KRUGER, C., PALACIO, D. and SUMMERS, M. (1991) 'Developing understanding of science concepts: Research-based in-service materials for primary teachers', *British Journal of In-service Education*, **17**, 3, pp.197–206.

LABBETT, B. (1995) 'Principles of procedure', in *Educational Action Research*, **3**, 1.

LAING, S. transl. (1930) *Snorre Sturlason Heimskringla: The Norse King Sagas*, London: Everyman ed.

LANG, S. (1992) *Times Educational Supplement*, History Extra, 11 December 1992, p.II.

LATHAM, R. transl. (1958) *The Travels of Marco Polo*, London: Folio Society ed.

LATHER, P. (1986) 'Research as praxis', *Harvard Educational Review*, **56**, 3, pp.257–77.

LAVENDER, R. (1975) *Myths, Legends and Lore*, Oxford: Blackwell.

LE GOFF, J. (1992) *History and Memory*, New York: Columbia University Press.

LEE, L. (1962) *Cider with Rosie*, London: Penguin Books.

LEE, D. (1974) transl. *Plato — The Republic*, London: Penguin.

LEE, P. (1978) 'Explanation and understanding in history', in DICKINSON, A. and LEE, P. *History Teaching and Historical Understanding*, London: Heinemann.

LEE, P. (1984) 'Historical imagination', in DICKINSON, A. (ed.) *Learning History*, London: Heinemann.

LEE, P. (1991) 'Historical knowledge and the national curriculum', in ALDRICH, R. (ed.) *History in the National Curriculum*, London: Kogan Page.

LEVI-STRAUSS, C. (1966) *The Savage Mind*, London.

LIGHTFOOT, L., HYMAS, C., HADFIELD, G. and SMITH, I. (1993) 'Sabotage of a reform', *Sunday Times*, 13 June, p.11.

LITTLE, V. (1983) 'What is historical imagination?', in *Teaching History*, **36**, pp.27–32.

LITTLE, V. and JOHN, T. (1990) *Historical Fiction in the Classroom*, HA Teaching of History Series 59, London: Historical Association.

LIVELY, A. (1991) *Times Educational Supplement*, 'Past into present', 22 November 1991, p.25.

LIVELY, P. (1987) *Moon Tiger*, London: Penguin.

LOW-BEER, A. (1967) 'Moral judgements in history and history teaching', in BURSTON, W. and THOMPSON, D. (eds) *Studies in the Nature and Teaching of History'*, London: Routledge.

LOYN, H. (1962) *Anglo-Saxon England and the Norman Conquest*, London: Longmans.

MACDONALD, F. and STARKEY, D. (1996) *Read Aloud History Stories*, London: Harper Collins.

MACINTYRE, A. (1981) *After Virtue*, London: Duckworth Press.

MADDERN, E. (1992) *Storytelling at Historic Sites*, London: English Heritage.

MANT, J. and SUMMERS, M. (1993) 'Some primary school teachers' understanding of the Earth's place in the universe', *Research Papers in Education*, **8**, 1, pp.101–29.

MARSDEN, W. (1993) 'Recycling religious instruction? Historical perspectives on contemporary cross-curricular issues', *History of Education*, **22**, 4, pp.321–33.

MASSEY, R. (1992) 'Expel Monty Python from history says enemy of trendy teaching', *Daily Mail*, 20 October.

MATTEN, J. (1984) *The Cult of St Edmund*, Pamphlet, Bury St Edmunds.

MATTHEWS, G. (1984) *Dialogues with Children*, London: Harvard University Press.

MAY, T. and WILLIAMS, S. (1987) 'Empathy, a case of Apathy?', *Teaching History*, October, pp.11–16.

MCALEAVY, T. (1994) 'Meeting pupils' learning needs', in BOURDILLON, H. (ed.) *Teaching History*, London: Routledge.

MCCABE, A. (1996) *Chameleon Readers: Teaching Children to Appreciate All Kinds of Good Stories*, New York: McGraw Hill.

MCCULLOCH, G. (1997) 'Privatising the past? History and education policy in the 1990s', *British Journal of Education Studies*, **45**, 1, pp.69–82.

MCEWAN, H. and EGAN, K. (1995) *Narrative in Teaching, Learning and Research*, New York: Teachers College Press.

MCKIERNAN, D. (1993) 'Imagining the nation at the end of the 20th Century', *Journal of Curriculum Studies*, **25**, 1, pp.33–51.

MEDLEY, R. and WHITE, C. (1991) 'Planning national curriculum assessment', in *History Teaching for Key Stage 3* Number 67, HA Teaching of History Series, London: Historical Association.

MEEK, M. (1991) *On Being Literate*, London: Bodley Head.

MEEK, M., WARLOW, A. and BARTON, G. (eds) (1977) *The Cool Web: The Pattern of Children's Reading*, London: The Bodley Head.

MIDDLETON, S. (1993) *Educating Feminists: Life Histories and Pedagogy*, New York: Teachers College Press.

MINK, L. (1981) 'Everyman his or her own analyst', in MITCHELL, W. (ed.) *On Narrative*, Chicago: University of Chicago Press.

MORRIS, C. (1992) 'Opening doors: Learning history through talk', in BOOTH, T. (ed.) *Learning for All: Curricula for Diversity in Education*, London: Routledge/Open University.

MORRIS, E. (1997) Speech to the School Curriculum and Assessment Authority's Conference on the Primary Curriculum, 10 June 1997.

MORRIS, W. (1882) in *Oxford Dictionary of Quotations*, 3rd edn., 1979, p.358.

MOSES, M. (ed.) (1994) *Stories from the Past*, Leamington Spa: Scholastic.

MURRAY, M. (1993) 'Little green lie', *Reader's Digest*, August, pp.53–57.

NAIDOO, B. (1992) *Through Whose Eyes?*, Stoke: Trentham Books.

NCC (1993) *Teaching History at KS1*, York: NCC Inset Resources.

NEEDHAM, G. (1966) *Aelfric Lives of Three English Saints*, London: Methuen.

NEELANDS, J. (1990) *Structuring Drama Work*, Cambridge: CUP.

NESBIT, E. (unknown) *Children's Stories from English History*, London: Raphael, Tuck and Sons Ltd.

NEUENSCHWANDER, J. (1976) *Oral History as a Teaching Approach*, New York: New York University Press.

NIAS, J. (1989) *Primary Teachers Talking — A Study of Teaching as Work*, London: Routledge.

NORQUAY, N. (1990) 'Life history research: Memory, schooling and social difference', in *Cambridge Journal of Education*, **20**, 3, pp.291–300.

O'DEA, J. (1994) 'Pursuing truth in narrative research', *Journal of Philosophy of Education*, **28**, 2, pp.161–71.

O'DONOGHUE, T. and SAVILLE, K. (1996) 'The power of the story form in curriculum development: An exposition on the ideas of Kieran Egan', *Curriculum*, **17**, 1, pp.24–35.

OFSTED (1993) *History at Key Stages 1, 2 and 3. Second year Final Report 1992–93*, London: HMSO.

OFSTED (1995) *The OFSTED Framework For The Inspection of Schools*, London: HMSO.

OFSTED (1997) *The Annual Report of Her Majesty's Chief Inspector of Schools*, London: OFSTED.

OFSTED (1998) *Standards in the Primary Curriculum 1996–97*, London: OFSTED.

ORCHARD, I. (1992) 'Oral history and teenagers', *Oral History The National Curriculum'*, **20**, 1, pp.58–63.

OSBORNE, J. and SIMON, S. (1996) 'Teacher subject knowledge: Implications for teaching and policy', Paper presented at BERA, September.

PALEY, V. (1997) *The Girl with The Brown Crayon*, Cambridge, MA: Harvard University Press.

PALSSON, H. (1971) *Hrafnkel's Saga and other Stories,* London: Penguin Classics.

PANKHANIA, J. (1994) *Liberating the National History Curriculum*, London: Falmer Press.

PAXTON, R. (1997) 'Someone with like a life wrote it: The effects of a visible author on High School history students', *Journal of Educational Psychology*, **89**, 2, pp.235–50.

PEACOCK, A. (1996) 'The role of text material in representing science knowledge to second language learners in primary schools', Paper presented at BERA, September.

PECK, S. (1992) 'Using oral evidence with infants: A toys and games project for 5–7 year olds', *Oral History*, **20**, pp.41–45.

PELLOWSKI, A. (1977) *The World of Storytelling*, New York: R.R.Bowker.

PENEFF, J. (1990) 'Myths in life histories', in SAMUEL, R. *The Myths We Live By*, London: Routledge.

PERERA, K. (1986) 'Some linguistic difficulties in school textbooks', in GILLHAM, B., *The Language of School Subjects*, London: Heinemann.

PERKS, R. (1992) *Oral History: Talking about the Past*, Historical Association, Pamphlet 94, 'Helps for students of history' series, London: Historical Association.

PERKS, R. and THOMSON, A. (eds) (1998) *The Oral History Reader*, London: Routledge.

PHILLIPS, M. (1994) 'The new tyranny destroying Britain', *Daily Mail*, 17 September, pp.8–9.

PHILLIPS, R. (1991) 'National curriculum history and teacher autonomy: The major challenge', *Teaching History*, **65**, October.

PHILLIPS, R. (1992) 'The battle for the big prize . . . school history and the national curriculum', *Curriculum Journal*, **3**, 3, pp.245–60.

POLANYI, M. (1983) *The Tacit Dimension*, London: Routledge and Paul.

POLANYI, M. and PROSCH, H. (1975) *Meaning*, Chicago: University of Chicago Press.

POLKINGHORNE, D. (1995) 'Narrative configuration in qualitative analysis', *Qualitative Studies in Education*, **8**, 1, pp.5–23.

POLLARD, A., BROADFOOT, P., OSBORN, M. and ABBOTT, D. (1994) *Changing English Primary Schools? The Impact of the Educational Reform Act at KS1*, London: Cassell.

POMPE, C. (1996) ' "But they're pink — who cares?": Popular culture in the primary years', in HILTON, M. (ed.) (1996) *Potent Fictions*, London: Routledge.

POPPER, K. (1992) 'Perspective: The magic of myths', *Times Higher Educational Supplement*, 24 July 1992, p.15.

PURKIS, S. (1980) *Oral History in Schools*, London: Oral History Society Publication.

RACKHAM, O. (1986) *The History of the Countryside*, London: J.M.Dent.

RADICE, B. transl. (1973) *The Letters of Abelard and Heloise*, London: Penguin Classics.

REDFERN, A. (1992) 'Both understanding and knowledge' oral history, *The National Curriculum*, **20**, 1, pp.29–33.

RICOEUR, P. (1981) 'Narrative time', in MITCHELL, W. (ed.) *On Narrative*, Chicago: University of Chicago Press.

RICOEUR, P. (1991) 'Life in quest of narrative', in WOOD, D. (ed.) *On Paul Ricoeur*, London: Routledge.

ROBERTS, M. (1990) 'History in the school curriculum 1972–1990: A possible dialectical sequence: Thesis, antithesis, synthesis?', *Curriculum Journal* **1**, 1, pp.65–75.

ROBINSON, J. and HAWPE, L. (1986) 'Narrative thinking as a heuristic process', in *Narrative Psychology*, SARBIN, T. (ed.) New York: Praeger.

ROBINSON, M. (1997) *Children Reading Print and Television*, London: Falmer Press.

ROEMER, K. (1983) 'Native American oral narratives: Context and continuity', in *Smoothing the Ground*, SWANN, B. (ed.) Berkeley: University of California Press.

ROGERS, P. (1984) 'Why teach history', in DICKINSON, A., LEE, P. and ROGERS, P. (eds) *Learning History*, London: Heinemann.

ROLLASON, D. (1989) *Saints and Relics in Anglo-Saxon England*, Oxford: Blackwell.

ROSEN, C. and ROSEN, H. (1973) *The Language of Primary School Children*, London: Penguin.

ROWLAND MARTIN, J. (1994) *Changing The Educational Landscape: Philosophy, Women and Curriculum*, London: Routledge.

SAMUEL, R. (1990) *The Guardian*, Education section, 13 March 1990.

SAMUEL, R. (1994) *Theatres of Memory*, London: Verso.

SARBIN, T. (ed.) (1986) *Narrative Psychology*, New York: Praeger.

SCAA (1997) *International Review of Curriculum and Assessment Frameworks*, London: SCAA/NFER.

SCARFE, N. (1969) 'The body of St Edmund', *Proceedings of the Suffolk Institute of Archaeology*, **31**, 3.

SCARFE, N. (1997) 'Jocelin of Brakelond's Identity: A review of the evidence', *Proceedings of the Suffolk Institute of Archaeology*, **39**, 1, pp.1–5.

SCHANK, R. (1990) *Tell Me a Story*, New York: Charles Scriber.

SCHOOLS COUNCIL (1969) *Humanities for the Young School Leaver: An Approach Through History*, London: Methuen.

SCOTT, B. (1994) 'A post-Dearing look at Hi.2: Interpretations of History', *Teaching History*, **75**, pp.20–6.

SELDON, A. and PAPWORTH, J. (1983) *By Word of Mouth — Elite Oral History,* London: Methuen.

SHAWYER, G., BOOTH, M. and BROWN, R. (1988) 'The development of children's historical thinking', *Cambridge Journal of Education*, **18**, 2, pp.209–19.

SHELLEY, M. (1990) *Telling Stories to Children,* Oxford: Thimble Press.

SHEMILT, D. (1984) 'Beauty and the philosopher', in DICKINSON, A. (ed.), *Learning History*, London: Heinemann.

SHERLEY-PRICE, L. transl. (1968) Bede, *A History of the English Church and People*, London: Penguin Classics.

SHULMAN, L. (1987) 'Knowledge and teaching: Foundations of the new reform', *Harvard Educational Review*, **57**, 1, pp.1–22.

SILVER, H. (1992) 'Knowing and not knowing', *History of Education*, **21**, 1, pp.97–108.

SINCLAIR, T. transl. (1962) 'Aristotle', *The Politics London*, London: Penguin Classics.

SLATER, J. (1991) 'History in the national curriculum: The final report of the history working group', in ALDRICH, R. (ed.) *History in the National Curriculum*, London: Kogan Page.

SMITH, B. '"That one horse" — making and reading stories across cultures', in STYLES, M., BEARNE, E. and WATSON, V. (eds) *Voices Off: Texts, Contexts and Readers*, London: Cassell.

SMITH, F. (1985) *Reading*, Cambridge: CUP, 2nd edn.

SMITH, F. (1991) *Times Educational Supplement*, 23 August 1991, 'In the company of authors', p.18.

SMITH, L. and SMITH, J. (1994) *Lives in Education*, New York: St Martin's Press.

SOBEL, D. (1995) *Longitude*, New York: Walker Publishing.

SOUTHGATE, B. (1996) *History: What & Why?*, London: Routledge.

SPENCE, D. (1986) 'Narrative smoothing and clinical wisdom', in SARBIN, T. (ed.) *Narrative Psychology*, New York: Praeger, pp.211–32.

STENHOUSE, L. (1967) *Culture and Education*, London: Heinemann.

STENHOUSE, L. (1975) *An Introduction to Curriculum Research and Development*, London: Heinemann.

STENHOUSE, L. (1978) 'Case study and case records: Towards a contemporary history of education', *British Educational Research Journal*, **4**, 2.

STENHOUSE, L. (1979) 'Research as a basis for teaching', in RUDDUCK, J. and HOPKINS, D. (eds) *Research as a Basis for Teaching*, London: Heinemann.

STENHOUSE, L. (1982) 'The process model in action: The HCP', in RUDDUCK, J. and HOPKINS, D. (eds) *Research as a Basis for Teaching*, London: Heinemann.

STUBBS, M. (1983) *Language, Schools and Classrooms*, London: Routledge.

SUFFOLK LEA (1991a) *Invaders and Settlers* — Guidance Booklet 10, Ipswich: Education Department.

SUFFOLK LEA (1991b) *Medieval Realms* — Guidance Booklet 11, Ipswich: Education Department.

SYLVESTER, D. (1994) 'Change and continuity in history teaching 1900–93', in BOURDILLON, H. *Teaching History*, London: Routledge.

TATE, N. (1995) 'The teaching of history and national identity' Speech, 18 September 1995.

TATE, N. (1996) 'Spiritual and moral aspects of the curriculum', Speech, 15 January 1996.

THATCHER, M. (1990) Parliamentary Debates (Hansard) 6th series, volume 170, 29 March, p.668.

THOMAS, D. (ed.) (1995) *Teachers' Stories*, Buckingham: Open University Press.

THOMAS, R. (1990) 'Ancient Greek family tradition and democracy', in SAMUEL, R. (1990) *The Myths We Live By*, London: Routledge.

THOMPSON, P. (1988) *The Voice of the Past*, Oxford 2nd ed: Oxford University Press.

TODD, R. (1981) 'Methodology: The hidden context of situation in studies of talk', in ADELMAN, C., *Uttering and Muttering*, London: Grant Macintyre.

TOMLINSON, S. (1990) *Multicultural Education in White Schools*, London: Batsford.

TONKIN, E. (1992) *Narrating Our Pasts: The Social Construction of Oral History*, Cambridge: Cambridge University Press.

TOOLAN, M. (1988) *Narrative: A Critical Linguistic Introduction*, London: Routledge.

TOUGH, J. (1979) *Talk for Teaching and Learning*, London: Ward Lock Educational.

TOWLER-EVANS (1997) 'It's not your everyday lesson is it?', in DAVIS, D. (ed.) *Interactive Research in Drama in Education*, Stoke: Trentham Books.

UNSWORTH, B. (1992) *Sacred Hunger*, London: Penguin.

VAN MANEN, M. (1991) *The Tact of Teaching*, New York: The Althouse Press.

VANSINA, J. (1980) 'Memory and oral tradition', in MILLER, J. (ed.) *The African Past Speaks*, Folkestone: Dawson.

VASS, P. (1993) 'Have I got a witness?', *Teaching History*, October, pp.19–25.

VICARY, T., HUGHES, P. and FAWCETT, V. (1993) *Oxford Primary History Scheme,* Oxford: Oxford University Press.

WACHTEL, N. (1971) *The Vision of the Vanquished*, Sussex: Harvester Press, 1977 edition.

WARNER, M. (1994) *From The Beast to the Blonde*, London: Chatto and Windus.

WARNER, R. transl. (1954) *Thucydides — The Pelopponesian War*, London: Penguin Classics.

WATKINS, T. (1992) 'Passing on history: A KS2 intergenerational project', in *Oral History 'The National Curriculum'*, Spring, **20**, 1, pp.47–53.

WEBSTER, G. (1993) *Boudica: The British Revolt Against Rome AD60*, London: Batsford.

WEDGWOOD, C.V. (1960) *Truth and Opinion*, London: Collins.

WEINER, G. (1993) 'Shell-shock or sisterhood: English school history and feminist practice', in ARNOT, M. (ed.) *Feminisim and Social Justice in Education*, London: Falmer Press.

WELLESLEY, K. transl. (1964) *Tacitus the Histories*, London: Penguin Classics.

WELLS, G. (1986) *The Meaning Makers*, London: Hodder and Stoughton.

WELLS, G. (1992) 'The centrality of talk in education', in NORMAN, K. (ed.) *Thinking Voices — The Work of the National Oracy Project*, London: Hodder and Stoughton, pp.283–310.

WHITE, H. (1981) 'The value of narrativity', in MITCHELL, W. (ed.) *On Narrative*, Chicago, IL: University of Chicago Press.

WHITE, H. (1987) *The Content of the Form*, New York: John Hopkins University Press.

WHITELOCK D. (1969) 'Fact and fiction in the legend of St Edmund', *Proceedings of the Suffolk Institute of Archaeology*, **31**, 3.

WHITELY, D. (1996) 'Reality in boxes: Children's perception of television narratives', in HILTON, M. (ed.) *Potent Fictions: Children's Literacy and the Challenge of Popular Cultures*, London: Routledge, pp.47–67.

WILCOX, B. (1997) 'Schooling, school improvement and the relevance of Alasdair MacIntyre', *Cambridge Journal of Education*, **27**, 2, pp.249–60.

WILKINSON, A. (1965) *Spoken English*, Birmingham: Educational Review Occasional Publications, No.2, Feb.

WILSON, M.D. (1985) *History for Pupils with Learning Difficulties*, London: Hodder and Stoughton.

WINSTON, J. (1996) 'Whose story? Whose culture? Moral and cultural values in Barbara Juster Esbensen's *The Star Maiden'*, *Children's Literature in Education*, **27**, 2, pp.109–21.

WISHART, E. (1986) 'Reading and understanding history textbooks', in *The Language of School Subjects*, GILHAM, B. ed. London: Heinemann.

WOODS, P. and JEFFREY, B. (1996) *Teachable Moments: The Art of Teaching in Primary Schools*, Buckingham: OUP.

WRAGG, E.C. (1993) *Primary Teaching Skills*, London: Routledge.

WRAY, D. and LEWIS, M. (1997) *Extending Literacy: Children Reading and Writing Non-fiction*, London: Routledge.

YOW, V.R. (1994) *Recording Oral History — A Practical Guide for Social Scientists,* London: Sage Publications.

ZIPES, J. (1995) *Creative Storytelling: Building Community, Changing Lives*, London: Routledge.

Index